SKILLS GAPS IN TWO MANUFACTURING SUBSECTORS IN SRI LANKA

Food and Beverages, and Electronics and Electricals

JANUARY 2020

ADB

ASIAN DEVELOPMENT BANK

© 2020 Asian Development Bank
6 ADB Avenue, Mandaluyong City, 1550 Metro Manila, Philippines
Tel +63 2 8632 4444; Fax +63 2 8636 2444
www.adb.org

Some rights reserved. Published in 2020.

ISBN 978-92-9261-910-7 (print), 978-92-9261-911-4 (electronic)
Publication Stock No. TCS190564-2
DOI: http://dx.doi.org/10.22617/TCS190564-2

The views expressed in this publication are those of the authors and do not necessarily reflect the views and policies of the Asian Development Bank (ADB) or its Board of Governors or the governments they represent.

ADB does not guarantee the accuracy of the data included in this publication and accepts no responsibility for any consequence of their use. The mention of specific companies or products of manufacturers does not imply that they are endorsed or recommended by ADB in preference to others of a similar nature that are not mentioned.

By making any designation of or reference to a particular territory or geographic area, or by using the term "country" in this document, ADB does not intend to make any judgments as to the legal or other status of any territory or area.

Corrigenda to ADB publications may be found at http://www.adb.org/publications/corrigenda.

Notes:
In this publication, "$" refers to United States dollars.
ADB recognizes "China" as the People's Republic of China and "Ceylon" as Sri Lanka.
All photos are by ADB unless otherwise stated.

Cover design by Noelito Francisco Trivino; photo on the left by ADB, and photo on the right by the University of Vocational Technology.

CONTENTS

TABLES AND FIGURES

TABLES

FIGURES

PREFACE

Sri Lanka ranks ahead of its peers in various indicators of basic education and health. Four decades after trade liberalization, Sri Lanka's manufacturing sector has not evolved, with a narrow set of products accounting for nearly two-thirds of the export basket in 2018. Exports of global value chain-linked products, which dominate trade and drive growth in many Asian economies, have not picked up in Sri Lanka. Addressing several constraints on investment would help attract new investment and, possibly, diversify the economic base. One such constraint is skills mismatch, in which the skills of workers produced by the education system fall short of the modern economy's needs.

This study examines the skills deficit in two subsectors—the food and beverages (FB) subsector, and the electronics and electricals (EE) subsector—and provides recommendations for addressing the gap. The study assesses the labor market to obtain a snapshot of the labor supply and compares it with labor demand expected in the FB and EE subsectors at different skill levels. The study finds that there will be significant unmet demand for skills in the FB subsector over the next 7 years. The skills gap in the EE subsector will also be large, in proportion to the size of the subsector. The technical and vocational education and training (TVET) sector faces several challenges such as absence of teacher training and industry exposure, poor recruitment procedures, inadequate resources, and lack of innovative changes. The TVET sector requires drastic reforms to meet the industry's rising skills demand, such as aligning curriculum with industry demand and continuous updating to keep up with technological innovation, modernizing management practices, and establishing public–private partnerships. Better coordination of development policies and programs among the key ministries and major institutions engaged in skills development is also needed.

Asian Development Bank (ADB) support for the TVET sector in Sri Lanka has focused on improving quality and relevance of skills development by introducing a competency-based training system, upgrading facilities and the capacity of relevant institutions, and strengthening nongovernment organization and public sector participation in the sector. ADB has expanded its area of assistance from the secondary sector to the tertiary sector. In 2018, ADB approved a project to support the Government of Sri Lanka's initiative to develop applied science and technology faculties in four universities and thereby produce technology-oriented graduates equipped with market-relevant skills. The findings of this study will inform ADB's future assistance to these sectors and will be a valuable input to the policy dialogue on skills development.

Sri Widowati, former country director of ADB's Sri Lanka Resident Mission, provided guidance and supported the study. Ramani Gunatilaka and Sunil Chandrasiri are the primary authors of the report. Utsav Kumar, senior country economist at ADB's Sri Lanka Resident Mission, led the study. Hasitha Wickremasinghe, senior economics officer at ADB's Sri Lanka Resident Mission, efficiently coordinated the project. Anouk Gowri Tyagarajah provided secretarial support throughout the study. Ma. Theresa Arago copyedited the document, while Joseph Manglicmot typeset it. Tuesday Soriano proofread the document and Noelito Francisco Trivino designed the cover.

Director general S. S. Mudalige of the Department of National Planning provided valuable guidance. Deputy director general W. J. L. A. Jayalath of the Tertiary and Vocational Education Commission (TVEC) and assistant director Nadeesh De Silva of the Tertiary and Vocational Education Commission provided data from the labor market surveys and helped respond to data-related queries. Director general I. R. Bandara of the Department of Census and Statistics supported the study by making unit-level data from the Labour Force Surveys and the Annual Survey of Industry available for the analysis. Senior officials of the Export Development Board shared data and reports related to the FB and EE subsectors. Executive director Champika Malalgoda of Sri Lanka's Board of Investment, and senior officials of the Board of Investment Priyanka Samaraweera and Ganga Palaketiya generously shared their data, research, and insights.

Ryotaro Hayashi, Uzma Hoque, Sudarshana Jayasundara, and Sakiko Tanaka of ADB; W. B. K. Bandara of the Sri Lanka Institute of Advanced Technological Education; and officers of the Government of Sri Lanka's Department of External Resources provided insightful comments, which helped sharpen the analysis and strengthen the policy recommendations.

Manjula Amerasinghe
Officer-in-Charge
Sri Lanka Resident Mission
South Asia Department
Asian Development Bank

ABBREVIATIONS AND CURRENCY EQUIVALENTS

ADB	Asian Development Bank
AL	advanced level
BOI	Board of Investment
DCS	Department of Census and Statistics
DTET	Department of Technical Education and Training
EDB	Export Development Board
EE	electronics and electricals
FB	food and beverages
ICT	information and communication technology
ILO	International Labour Organization
ISIC	International Standard Industrial Classification
LDS	Labour Demand Survey
LFS	Labour Force Survey
MICRCDVS	Ministry of Industry and Commerce, Resettlement of Protracted Displaced Persons, Co-operative Development and Vocational Training and Skills Development
NAITA	National Apprentice and Industrial Training Authority
NCECP	National Certificate in Electronic Crafts Practice
NES	National Export Strategy
NVQ	National Vocational Qualification
NYSC	National Youth Services Council
OJT	on-the-job training
OL	ordinary level
STEP	Skills Toward Employment and Productivity
TOT	training of trainers
TVEC	Tertiary and Vocational Education Commission
TVET	technical and vocational education and training
VTA	Vocational Training Authority

CURRENCY EQUIVALENTS
(as of 26 November 2019)

Currency unit	–	Sri Lanka rupee/s (SLRe/SLRs)
SLRe1.00	=	$0.0055
$1.00	=	SLRs183.26

EXECUTIVE SUMMARY

Persistent skills gaps in priority growth sectors continue to impede economic growth and the creation of decent jobs in Sri Lanka. A major part of the problem lies with poor student learning achievements in the general education sector cycle, particularly in core subjects such as mathematics, science, and the English language. Weaknesses in the technical and vocational education and training (TVET) sector, including a lack of alignment with the market needs and problems with management and coordination, have also impeded the supply of work-oriented skills. While successive policy regimes have recognized the need to reduce skills gaps in construction, tourism, light engineering, and information and communication technology, skills gaps in other sectors with the potential for export growth and promising rates of labor absorption have received less attention. This study is designed to address the issue of skills gaps in two manufacturing subsectors: food and beverages (FB), and electronics and electricals (EE). Both have been identified as priority sectors for promoting export-led growth by the Board of Investment and the Export Development Board.

Key Findings

(i) Projected demand for labor in the FB subsector is expected to increase from 316,323 in 2019 to 439,827 by 2025 under pessimistic growth assumptions. Under optimistic assumptions, the projected increase is to 554,162 during the same period.

(ii) Projected demand for labor in the EE subsector is expected to increase from 18,601 in 2018 to 26,161 by 2025 under pessimistic growth assumptions. Under optimistic assumptions, the projected increase is to 31,102 during the same period.

(iii) Graduate outputs (i.e., the number of students graduating from all courses) are 1,766 in the FB subsector and 706 in the EE subsector in 2019. These estimates are expected to rise to 2,576 (FB subsector) and 947 (EE subsector) by 2025. Female participation in training targeted at the FB subsector is 53% and at the EE subsector, 3%, as against a national average of 41% female participation in the TVET sector.

(iv) The dropout rates (i.e., the percentage of students that do not complete the course out of the total number enrolled) are 51% (FB subsector) and 61% (EE subsector) as against the TVET sector average of 31%.

(v) The skills gap (i.e., the shortfall of supply over demand) in the FB subsector is expected to vary from 27,110 in 2019 to 18,132 by 2025 under pessimistic estimates. The expected change in skills gaps in the EE subsector is from 996 to 857 for the same period.

(vi) The skills gap in the FB subsector is expected to increase from 27,110 in 2019 to 51,603 by 2025 under optimistic estimates. The expected increase in skills gaps in the EE subsector is from 996 to 2,319 for the same period.

Deficits in core competencies and technical skills among high-skilled workers are weaker in the FB subsector compared with the EE subsector. Among medium-skilled occupations in the two subsectors, workers in the EE subsector lack several core competencies. Recommendations for action are as follows:

(i)   Conduct yearly campaigns in schools to trigger interest in the EE and FB subsectors on employment potential and career development prospects.

(ii)   Introduce an industry–educational/training institution coordination exchange mechanism by organizing short-term contracts for lecturers/instructors to serve as consultants or gain some knowledge on technological applications in the EE and FB subsectors.

(iii) Develop short- and medium-term courses and on-the-job training facilities through training providers, focusing on skills identified as lacking through a skills gap assessment.

(iv)   Revise course curricula of EE and FB course programs offered by training providers in consultation with the industry.

(v)   Conduct training-of-trainers program for EE and FB subsector instructors jointly by the industry and the universities.

(vi)   Provide industry experience and exposure to training instructors/lecturers attached to training institutes in consultation with the industry.

(vii) Provide industry experience and exposure to trainees who follow training courses offered by the training institutes.

(viii) Provide occupation-specific on-the-job training facilities to trainees enrolled in EE- and FB-related courses in consultation with the industry.

(ix)   Revise course evaluation methodologies of EE- and FB-related courses in consultation with the industry.

(x)   Establish separate skills councils for the EE and FB subsectors.

(xi)   Create a mobile platform to offer some course modules of EE and FB study programs.

(xii) Diversify program mix of the FB and EE subsectors through the Employment-Linked Training Purchasing model implemented by the Skills Sector Development Programme.

(xiii) Address the issue of long delays in issuing National Vocational Qualification certificates.

(xiv) Conduct skills gaps study based on business establishments.

(xv)   Introduce the performance-based allowance scheme for the instructors and supervisors of the TVET sector.

CHAPTER 1
INTRODUCTION

1.1   Rationale and Objectives

Although Sri Lanka was the first country in South Asia to liberalize its economy and emphasize export-led growth in its policy framework, objectives of structural transformation and export diversification have been met only partially over the last 4 decades. While the nearly 30-year civil conflict acted as a constraint during much of the period, the decade following the end of the conflict saw little change in the country's export portfolio. Exports, as a share of gross domestic product (GDP), fell from 33.3% in 2000 to 12.5% in 2016 before increasing slightly to 13.4% in 2018. Government expenditure on infrastructure became the main impetus for economic growth (Weerakoon, Kumar, and Dime 2019; World Bank 2019). Meanwhile, the incidence of poverty declined and real wages rose, especially among the low-skilled, which may have been partly due to demographic change and the tightening labor market rather than any appreciable increase in productivity growth (Chandrasiri, De Mel, and Jayathunge 2017). Nearly two-thirds of Sri Lanka's employed labor force still work under informal conditions with the pace of informal job growth far outstripping job creation in formal employment (Majid and Gunatilaka 2017).

Sri Lanka has not been able to fully realize its potential in export-led growth for many reasons. These include access to land, skills gap, restrictive product market regulations, inward-oriented trade policies which created an "anti-export" bias, a weak business climate, and complex tax regime prior to enactment of the new Inland Revenue Act in April 2018. Athukorala et al. (2017), Center for International Development (2018), and World Bank (2015) are some of the studies that discuss the constraints on Sri Lanka's growth prospects. Many institutional factors also constrain growth, several relating to the country's political economy. Key among them is the lack of necessary institutional and policy support to generate the skills essential to upgrade and diversify the economy's production and export bases. To generate sources of sustained foreign exchange earnings, major structural reforms are needed, particularly those that prioritize export-led growth and diversification. These reforms would include structurally transforming Sri Lanka's human capital in line with technologically driven market demand. While Sri Lanka's demographic window has ended and its population is aging, the postwar baby boom (de Silva and de Silva 2015) may provide a narrow window of opportunity over the next decade to accelerate economic growth and incomes, so long as young workers entering the labor force have the skills to engage in productive work.

In line with informing the formulation of appropriate skills development policies, this study uses the most recently available sample survey and administrative data to assess skills gaps in two manufacturing subsectors that have the potential for export-led growth and capacity for high levels of labor absorption. The subsectors were selected according to a series of criteria described in Chapter 2. The selection process found the food and beverages, and electronics and electricals subsectors as needing to be prioritized for targeted skills development. The next section describes the data and methods used for the analysis.

1.2   Data and Methods

Definition of skills

The present study is industry specific. Hence, the assessment of skills gaps is based on core competencies and technical skills. Core competencies help in the development of core products and enhance the competitiveness of a firm. In broad terms, they comprise a set of human skills acquired via teaching, reading, or interaction with others in real life. Core competencies, behaviors, and attributes enable a person to function in the workplace. While different terms are used in different countries to define them, these skills typically include communication skills, problem-solving, computer literacy, teamwork, and time management.[1] In training programs, such skills are usually imparted as modules and are assessed. The common core competency skills include communication, problem-solving, integrity, stability, interpersonal skills, extroversion, and openness. However, effective application of these skills is not only conditioned by training in them but also by the presence of appropriate personality traits, behaviors, and attitudes in a workplace.

Technical skills are used to describe occupation-specific competencies which demonstrate proficiency or mastery over technical know-how and their application to given work assignments/processes.[2] Technical skills are the abilities and knowledge needed to perform specific tasks. They are practical, and often relate to mechanical processes, information technology, mathematical operations, or scientific tasks. They are more job-specific than core competencies and are measured in terms of technology use, computer use, mechanical use, machinery use, English language ability, the ability to work autonomously, and manual labor skills. Technical skills are acquired through learning, training, and on-the-job training (OJT). This approach of skills gap assessment is consistent with the technical and vocational education and training (TVET) sector skills development programs.

Data and methods

Skills gaps are the shortfall of supply over demand and, in this study, skills gaps are estimated by first estimating the demand for skills, and then their supply, and thereafter matching the two. Due to lack of a longitudinal data set with information on the factors conditioning labor demand necessary to estimate a model of skills demand or supply, information from a variety of sources is collated. To estimate demand, data from the Labour Demand Survey of 2017, Quarterly Labour Force Surveys of 2016, and the Annual Survey of Industries 2011–2015 of the Sri Lanka Department of Census and Statistics (DCS) as well as 2012 data from the Skills Toward Employment and Productivity (STEP) initiative of the World Bank are used. To estimate supply, administrative data from training providers, particularly the Tertiary and Vocational Education Commission (TVEC) and affiliated institutes, are used. The analysis is also informed by qualitative data gathered from consultations with stakeholders, that is, employers and training providers, and analyzed in the context of skills supply and skills gaps.

The methodological approach consisted of several steps. First, growth employment elasticities are estimated using data from the DCS Annual Survey of Industries 2011–2015 and baseline employment figures by occupation from the DCS Labour Demand Survey of 2017, to project the demand for labor by occupation from 2019 to 2025 (Chapter 3). Additionally, the chapter also estimates deficits in core competencies and technical skills using the World Bank's STEP data and the DCS Quarterly Labour Force Survey of 2016. In Chapter 4, administrative data on recent actual intake figures and dropout figures in TVET and higher educational institutes are used to estimate expected output of graduates over the period 2019–2025 which are then matched with the labor demand projections to estimate skills gaps.

---

[1]   For more details, see Lathi (1999); Gallon, Stillman, and Coates (1995); Esque and Gilbert (1995); and Prahalad and Hamel (1990).

[2]   For more details, see Dundar et al. (2014) and references cited therein.

The next chapter provides extensive information about the background of the study, beginning with an overview of Sri Lanka's labor market and an assessment of the skills-generating capacity and performance of the general education and TVET sectors. This is followed by an account of the criteria used to select the two manufacturing subsectors for the skills gaps analysis and a review of the skills-related characteristics of the two subsectors. Chapter 3 focuses on estimating the demand for skills in the two subsectors, while Chapter 4 looks at supply and estimates the skills gaps. Chapter 5 concludes with an overview of the findings and detailed recommendations for policy action.

CHAPTER 2

BACKGROUND

2.1   Overview of Sri Lanka's Labor Market

Sri Lanka posted a relatively respectable average growth rate of 5.6% from 2010 to 2018, but has had only limited change in the structural composition of output (Figure 2.1). Agriculture's contribution to gross value added (at basic prices), at 9.5% in 2010, declined to 8.6% in 2018, while the industry sector's contribution has remained stable, at 29.7% in 2010 and 29.4% 2018. In contrast, the contribution to gross value added of Sri Lanka's services sector increased marginally from 60.9% in 2010 to 62.0% in 2018. The structure of employment has changed more noticeably, but not dramatically. For example, over the same period, the employment share of the services sector rose from 40% to 45%, the share of agriculture declined from 34% to 26%, and the share of the industry sector inched up from 26% to 28%. The following sections take a closer look at the characteristics of Sri Lanka's labor market.

Demographic transition and participation

Sri Lanka's population has been rapidly aging with share of working-age population in total population declining (Figure 2.2) and the absolute number of working-age population projected to decline after 2027 based on United Nations population projections. However, since fertility rates increased after the civil conflict ended in 2009, a "youth bulge" is likely to emerge as the youth population (15–29 years old) is expected to increase (Figure 2.3) (de Silva and de Silva 2015). However, this bulge may not be enough to stem the decline in working-age population. Yet it does provide Sri Lanka another chance at stimulating growth if, over the next 10 years, the correct mix of policies is put in place. Foremost among them are policies and programs to forge a highly educated and skilled workforce, facilitate technology transfer, improve the investment climate, and provide institutional and other supportive systems (de Silva and de Silva 2015). This would require a paradigm shift in the generation of skills and concerted efforts at reducing the skills gap.

Meanwhile, as Sri Lanka's population ages, its labor market is becoming tight. The male participation rate remains high and is unlikely to increase further, while the female labor force participation rate continues to remain low as women appear unwilling or unable to engage in paid work (Figure 2.4). While women's participation rate has been low and stable, there were some variations by sector over the longer period, from 1996 to 2016 (Figure 2.5). Participation rates have been, historically, highest in the estates sector, though these show a long-term decline. Women in rural areas have marginally increased their participation, but urban women hardly at all. While labor force participation rates at the national level have been remarkably stable, unemployment rates have declined (Figure 2.4). The decline in the unemployment rate and a rise in the employment-to-population ratio underlie the stability in participation.

Figure 2.1: Trends in the Structure of Output and Employment in Sri Lanka, 2010–2018

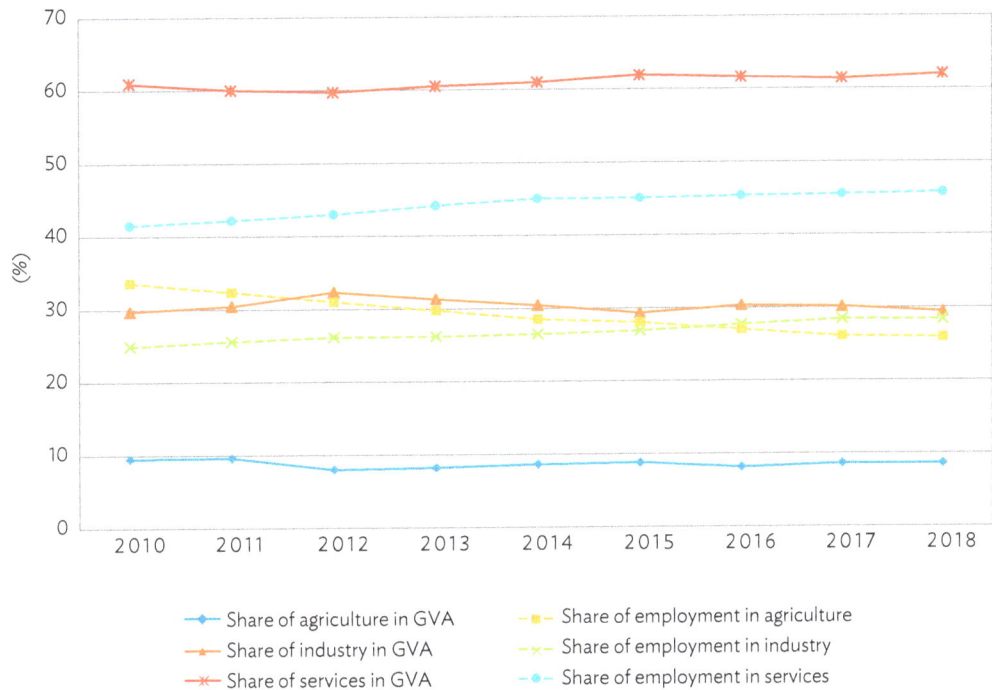

GVA = gross value added.
Sources: Central Bank of Sri Lanka. Annual Reports (various years); and World Bank. World Development Indicators (accessed 14 October 2018).

Figure 2.2: Share of Elderly and Working-Age Population in Total Population, 2015

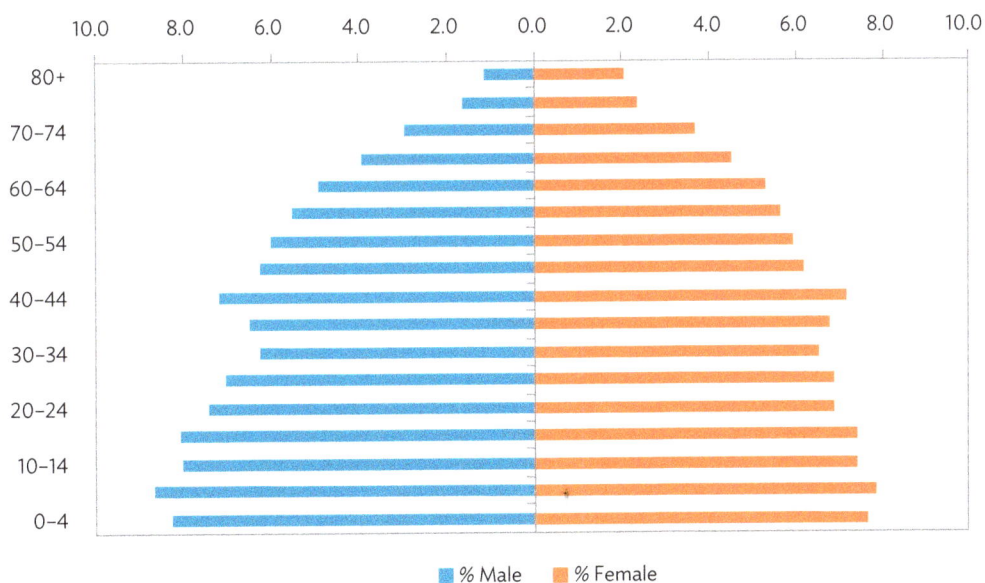

Note: Working-age population comprise those in the age group 15–59 years, while the elderly are those aged 60 years and above.
Source: de Silva and de Silva (2015).

Figure 2.3: Sri Lanka's Youth Population, 1971–2047

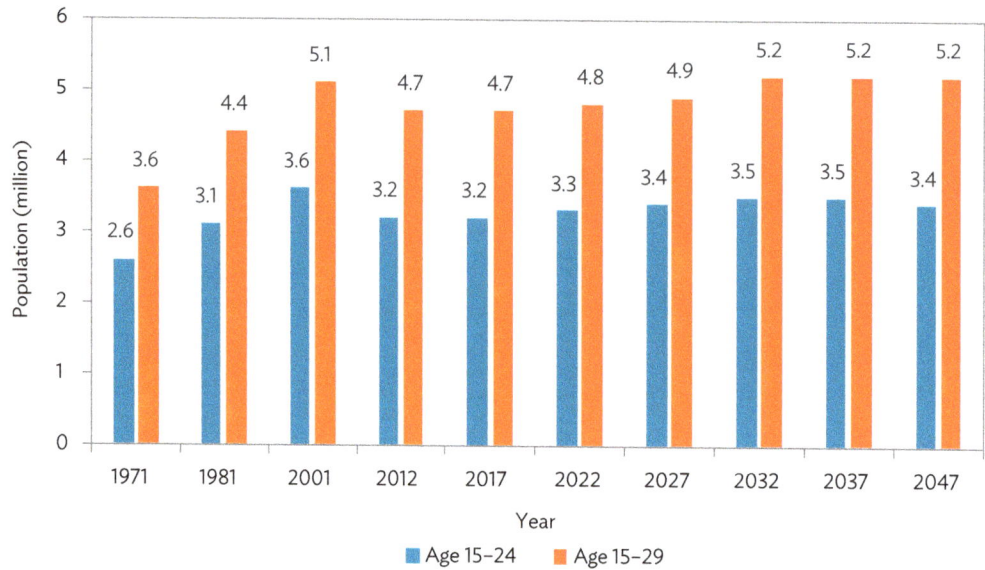

Note: Data for 2017–2047 are based on population projection by de Silva and de Silva (2015).
Source: de Silva and de Silva (2015).

Figure 2.4: Trends in Labor Force Participation and Unemployment in Sri Lanka by Gender, 2006 2017

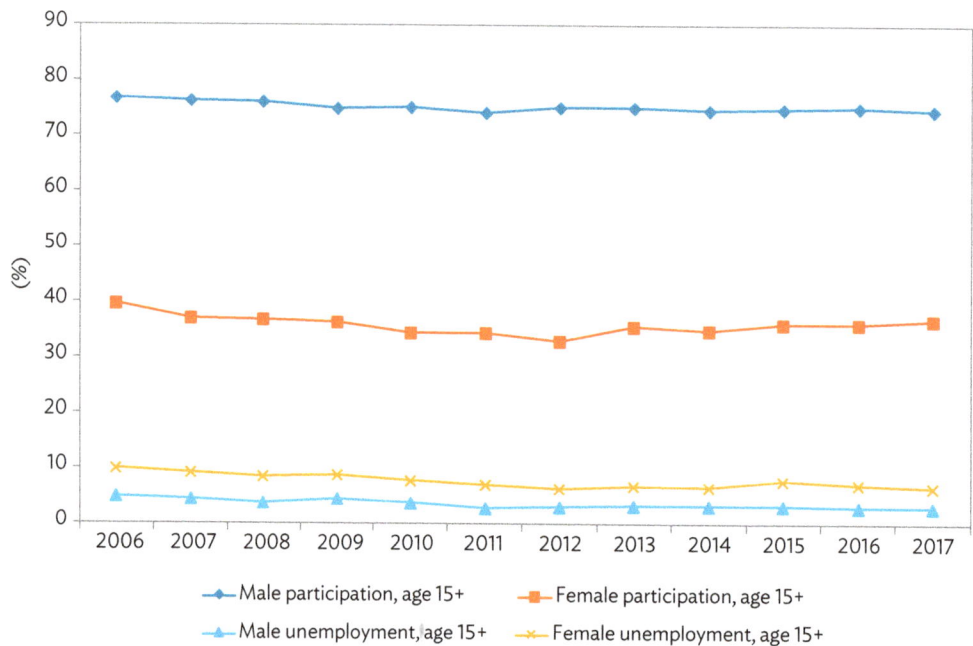

Notes: Data for 2009 onward include all districts. Data for previous years either excluded both Northern Province and Eastern Province, Northern Province only, or some districts of Northern Province. For details, see Department of Census and Statistics (2018).
Sources: Department of Census and Statistics (2015 and 2018).

**Figure 2.5: Trends in Women's Labor Force Participation
in Sri Lanka by Sector, 1996–2016**

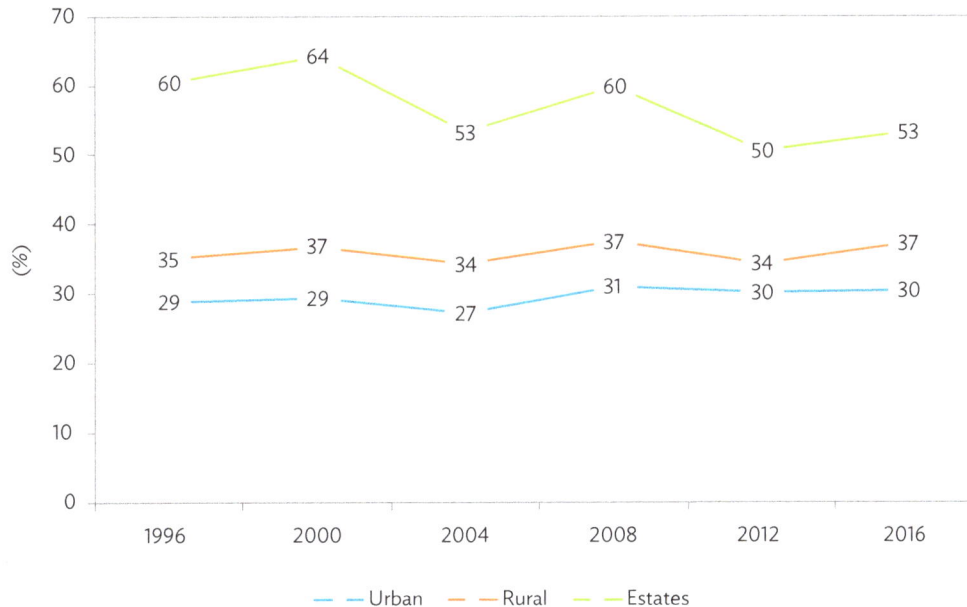

Sources: Data for 1996 to 2008 are from Gunatilaka (2013). Data for 2012 and 2016 are estimated from Department of Census and Statistics Quarterly Labour Force Surveys for those years.

Unemployment and underemployment

Low unemployment rates mask several features of unemployment (Dundar et al. 2014). First, youth unemployment rates are very high (18.5% in 2017 for the 15–24 age cohort) especially among women: at least a fourth of the female youth population is seeking paid work (DCS 2018). Half of all those looking for work have been educated at least up to General Certificate of Education (GCE) ordinary level (OL) and 32% of unemployed people have got through their GCE advanced level (AL) examination (Figure 2.6). Second, low unemployment disguises underemployment. When defined as those individuals who have worked less than 35 hours a week in both main and secondary occupations and are prepared and available for further paid work if provided, 2.2% of employed men and 3.9% of employed women were found to be underemployed in 2017. Agriculture appears to be the repository of underemployment at 4.9% of the workforce compared with industry (2.8%) and services (1.6%) (DCS 2018).

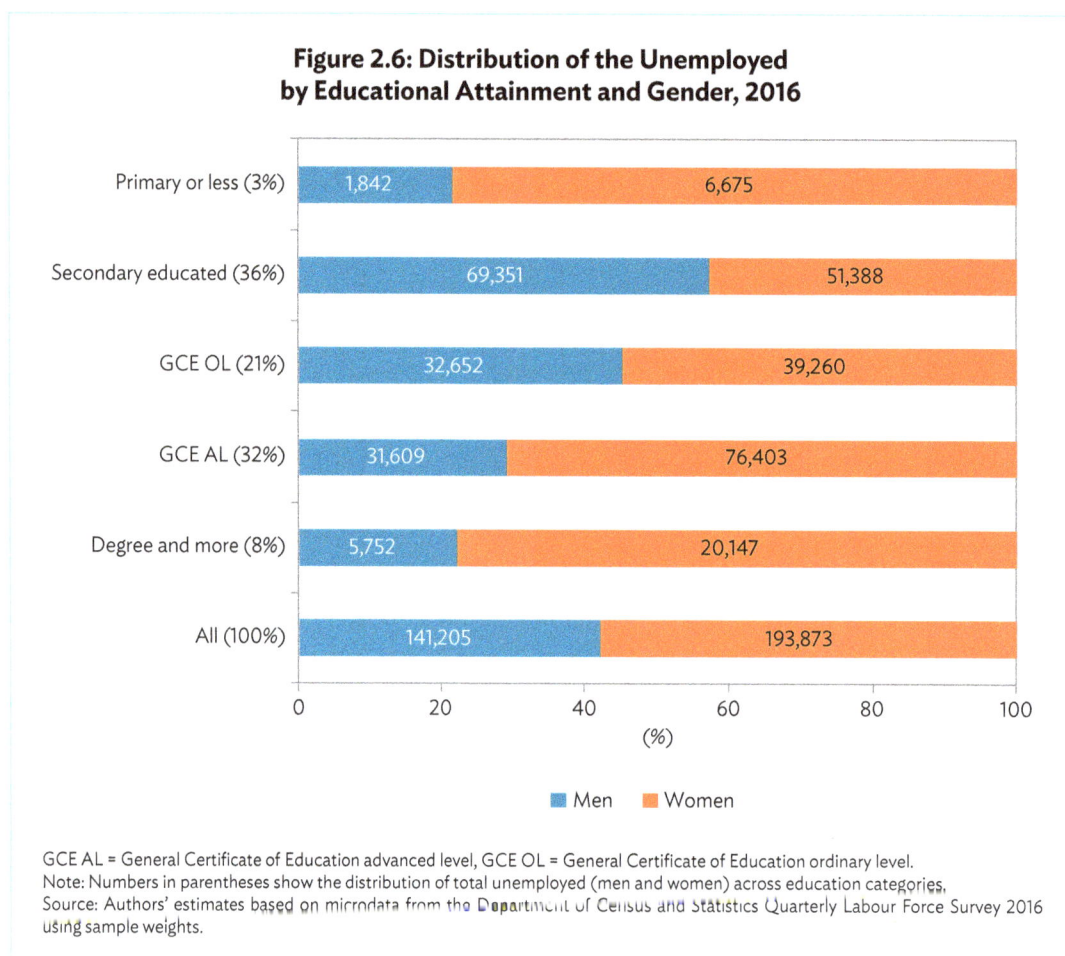

**Figure 2.6: Distribution of the Unemployed
by Educational Attainment and Gender, 2016**

Education category	Men	Women
Primary or less (3%)	1,842	6,675
Secondary educated (36%)	69,351	51,388
GCE OL (21%)	32,652	39,260
GCE AL (32%)	31,609	76,403
Degree and more (8%)	5,752	20,147
All (100%)	141,205	193,873

(%)

■ Men   ■ Women

GCE AL = General Certificate of Education advanced level, GCE OL = General Certificate of Education ordinary level.
Note: Numbers in parentheses show the distribution of total unemployed (men and women) across education categories.
Source: Authors' estimates based on microdata from the Department of Census and Statistics Quarterly Labour Force Survey 2016 using sample weights.

Employment

There were nearly 16 million Sri Lankans of working age in 2017. Among them, 8.5 million were employed (DCS 2018). The distribution of the employed workforce across job status categories hardly changed between 2006 and 2017. The public sector still accounts for half of the formal sector (Figure 2.7). The share of public employment in total employment increased from 13% to 15% between 2006 and 2016 (Ministry of Finance [various years]). Nor has the share of own-account workers (around 33%) changed over the period. Private informal employment accounted for a little more than a fourth of all those working for pay.

Figure 2.8 shows the distribution of employment across gender. While agriculture (including fishing and forestry) employed the largest share (29%) of the workforce in 2016, 40% of such jobs were held by women. Manufacturing employed 18% of workers, and here, women's share was slightly more than half. All other sectors were dominated by men, especially the large wholesale and retail trading sector which employed 14% of the total workforce, and construction, which was the fourth-largest sector accounting for nearly 7% of all jobs. Women were predominant in the health and education sectors.

Figure 2.7: Job Status of the Employed, 2016

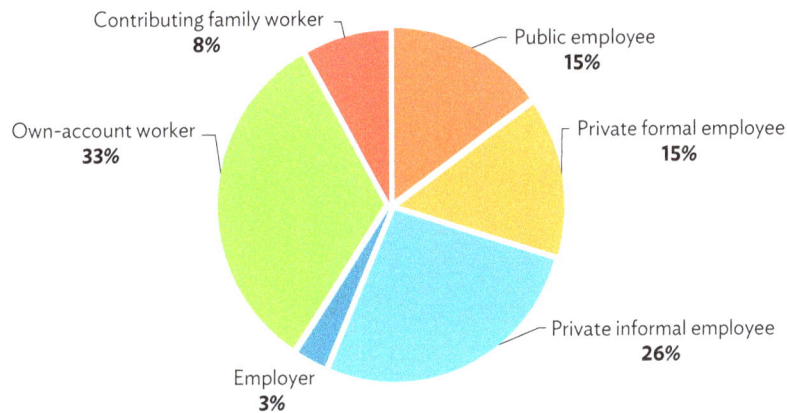

- Contributing family worker **8%**
- Public employee **15%**
- Own-account worker **33%**
- Private formal employee **15%**
- Private informal employee **26%**
- Employer **3%**

Source: Authors' estimates based on microdata from the Department of Census and Statistics Quarterly Labour Force Survey 2016 using sample weights.

Figure 2.8: Distribution of Employment across Economic Sectors by Gender, 2016

Economic Sectors

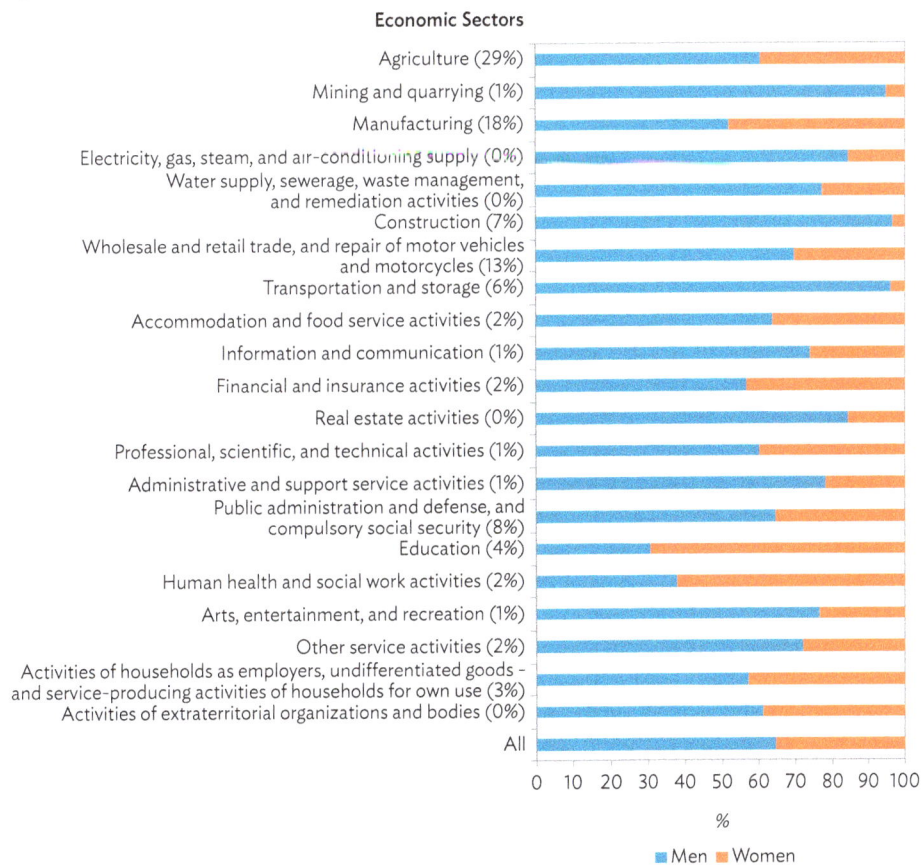

- Agriculture (29%)
- Mining and quarrying (1%)
- Manufacturing (18%)
- Electricity, gas, steam, and air-conditioning supply (0%)
- Water supply, sewerage, waste management, and remediation activities (0%)
- Construction (7%)
- Wholesale and retail trade, and repair of motor vehicles and motorcycles (13%)
- Transportation and storage (6%)
- Accommodation and food service activities (2%)
- Information and communication (1%)
- Financial and insurance activities (2%)
- Real estate activities (0%)
- Professional, scientific, and technical activities (1%)
- Administrative and support service activities (1%)
- Public administration and defense, and compulsory social security (8%)
- Education (4%)
- Human health and social work activities (2%)
- Arts, entertainment, and recreation (1%)
- Other service activities (2%)
- Activities of households as employers, undifferentiated goods - and service-producing activities of households for own use (3%)
- Activities of extraterritorial organizations and bodies (0%)
- All

■ Men ■ Women

Note: Numbers in parentheses show shares of total employment by sector.
Source: Authors' estimates based on microdata from the Department of Census and Statistics Quarterly Labour Force Survey 2016 using sample weights.

A fourth of all Sri Lankans employed in the private sector work in large firms with more than 100 employees (Figure 2.9). A little more than that work in microenterprises that are predominantly informal and with fewer than five employees. Small and medium-sized firms' share of total employment account for a little more than a fourth.

Figure 2.9: Distribution of Employment across Firm Size by Gender, 2016

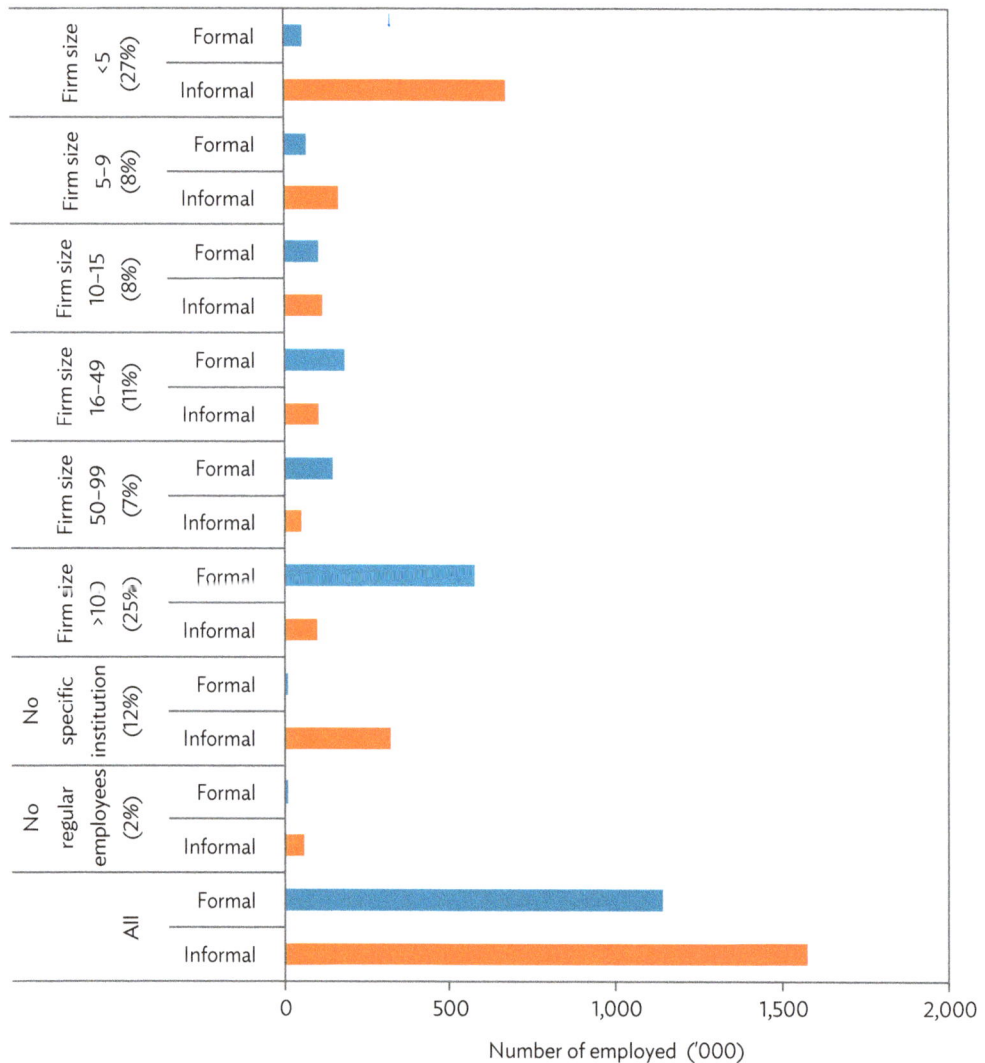

Note: Numbers in parentheses show shares of total employment by size category of firms.
Source: Authors' estimates based on microdata from the Department of Census and Statistics Quarterly Labour Force Survey 2016 using sample weights.

Occupations and skills

Figure 2.10 sets out the distribution of workers across occupations according to gender. High-skilled workers (managers, professionals, and technicians and associated workers) account for a fifth of all workers, and low-skilled workers (craft workers, machine operators and assemblers, and elementary workers) for nearly half. Women account for a little more than half of all clerical workers but only for 14% of machine operators. Although workers' educational attainment has improved over the years (Majid and Gunatilaka 2017), those with only lower-secondary education account for nearly half of the workers (Figure 2.11). It is apparent that even while demographic transition squeezes the supply of workers, most of them do not have the capacity to work in jobs that require higher levels of skills. Previous studies on Sri Lanka emphasize the constraints that the lack of noncognitive skills and technical skills relating to English and information and communication technology (ICT) place on economic expansion and diversification (for example, Dundar et al. 2014). Table 2.1 shows the distribution of men and women workers by the highest level of education attained across occupations. Meanwhile, Gunewardena's (2015) analysis of gender-based returns to noncognitive skills based on the World Bank's STEP 2012 data shows that, although women on average have higher levels of noncognitive skills than men, the wage system does not reward them for those skills.

Figure 2.10: Distribution of Employment across Occupations and Gender, 2016

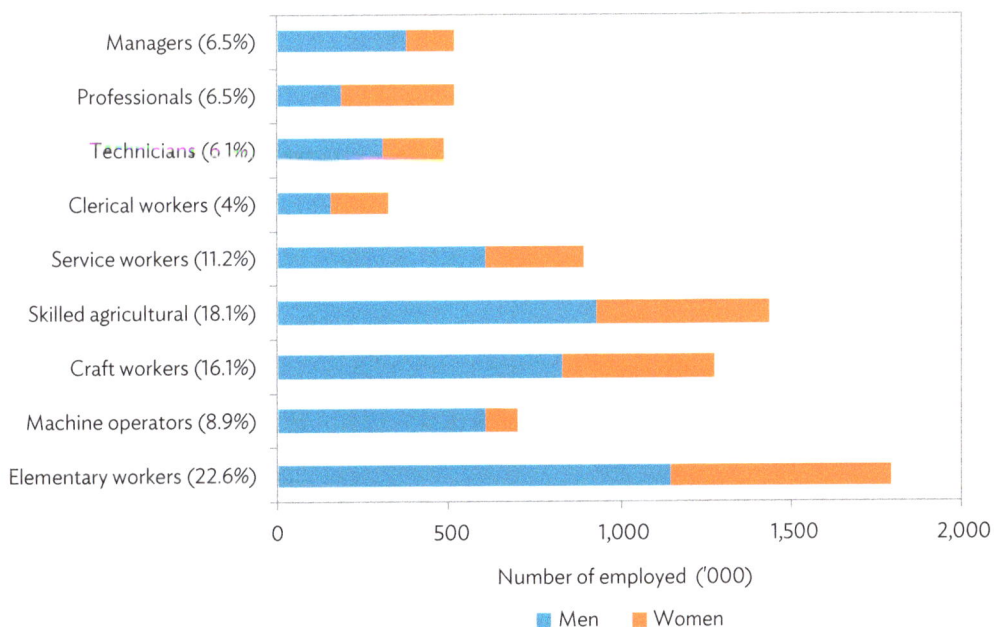

Note: Numbers in parentheses show shares of total employment by occupation.
Source: Authors' estimates based on microdata from the Department of Census and Statistics Quarterly Labour Force Survey 2016 using sample weights.

Figure 2.11: Distribution of Employment across Education Categories and Gender, 2016

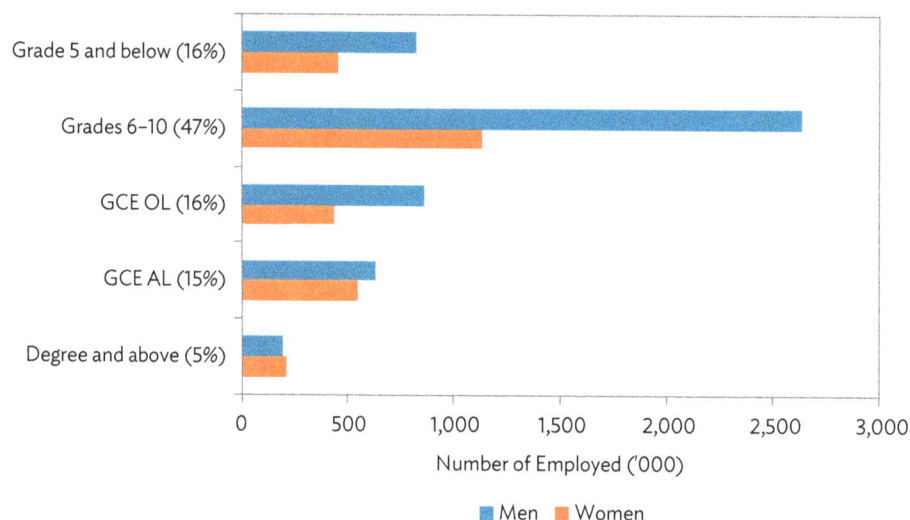

GCE AL = General Certificate of Education advanced level, GCE OL = General Certificate of Education ordinary level.
Note: Numbers in parentheses show shares of total employed by highest level of education attained.
Source: Authors' estimates based on microdata from the Department of Census and Statistics Quarterly Labour Force Survey 2016 using sample weights.

Table 2.1: Education Level by Gender and Occupation (%), 2016

Occupation	Educational Attainment Category	Men	Women
Managers	Grade 5 and below (6.5%)	6.3	7.0
	Grades 6–10 (34.1%)	34.0	34.5
	GCE OL (22.8%)	23.7	20.4
	GCE AL (26.6%)	27.0	25.5
	Degree and above (10%)	9.0	12.7
Professionals	Grade 5 and below (0.80%)	1.7	0.2
	Grades 6–10 (6.9%)	8.5	6.0
	GCE OL (12.9%)	11.8	13.5
	GCE AL (40.9%)	32.6	45.7
	Degree and above (38.5%)	45.4	34.6
Technicians	Grade 5 and below (1.1%)	1.1	1.1
	Grades 6–10 (15.5%)	19.0	9.2
	GCE OL (24.2%)	28.0	17.5
	GCE AL (39.5%)	37.7	42.8
	Degree and above (19.6%)	14.2	29.4

continued on next page

Table 2.1 continued

Occupation	Educational Attainment Category	Men	Women
Clerical workers	Grade 5 and below (0.50%)	0.8	0.3
	Grades 6–10 (13.5%)	19.6	7.7
	GCE OL (22.3%)	26.1	18.7
	GCE AL (54.4%)	48.8	59.6
	Degree and above (9.3%)	4.7	13.7
Service workers	Grade 5 and below (9.6%)	9.9	9.1
	Grades 6–10 (48.2%)	50.8	42.5
	GCE OL (22.7%)	22.5	23.0
	GCE AL (18.6%)	15.9	24.3
	Degree and above (0.90%)	0.8	1.1
Skilled agricultural workers	Grade 5 and below (26.6%)	28.2	23.5
	Grades 6–10 (53.5%)	53.5	53.5
	GCE OL (13.1%)	12.0	15.1
	GCE AL (6.3%)	5.5	7.8
	Degree and above (0.50%)	0.7	0.2
Craft workers	Grade 5 and below (12.2%)	11.8	12.8
	Grades 6–10 (62.8%)	64.9	58.8
	GCE OL (17.2%)	16.7	18.3
	GCE AL (7.4%)	6.2	9.5
	Degree and above (0.50%)	0.4	0.6
Machine operators and assemblers	Grade 5 and below (5.9%)	5.9	5.8
	Grades 6–10 (64.2%)	64.2	64.2
	GCE OL (21.9%)	22.5	18.2
	GCE AL (7.7%)	7.1	11.6
	Degree and above (0.30%)	0.3	0.1
Elementary workers	Grade 5 and below (31.9%)	29.7	35.6
	Grades 6–10 (54.9%)	58.0	49.5
	GCE OL (8.9%)	8.6	9.3
	GCE AL (4.2%)	3.5	5.3
	Degree and above (0.20%)	0.2	0.3

GCE AL = General Certificate of Education advanced level, GCE OL = General Certificate of Education ordinary level.
Notes: Numbers in parentheses show shares of total employed by highest level of education attained within each occupation category. Data in each column show distribution within each occupation by highest level of education for males and females separately.
Source: Authors' estimates based on microdata from the Department of Census and Statistics Quarterly Labour Force Survey 2016 using sample weights.

2.2.   Skills-Generating Capacity of the General Education and TVET Sectors

Sri Lanka's general education cycle comprises four main stages: (i) primary (Grades 1–5); (ii) junior secondary (Grades 6–9); (iii) senior secondary or GCE ordinary level (OL), Grades 10–11; and (iv) GCE advanced level (AL), Grades 12–13.[3] There is a common syllabus including both compulsory and optional subjects for the OL.[4] At AL, students are free to select one stream out of four: Arts, Science, Technology, and Commerce.[5] English and General Knowledge are compulsory subjects for AL students. The tertiary education system consists of technical and vocational education and training (TVET) and university education. Both TVET and university education are dominated by the government sector, but the nongovernment sector's participation has been expanding over the last 15 years due to capacity limitations of state sector service providers and high demand. For example, at present, the TVET sector comprises 1,147 service providers representing the public (46%), and the private and nongovernment sectors (56%).[6] However, leading public sector TVET institutes operate through an island-wide network of training institutes and account for 77% of total enrollment and 74% of TVET graduate output. At present, enrollment is around 302,000 students per year at Grade 1 level, around 189,000 at the TVET level, and around 40,000 at the university level (Appendix, Figure A.1). The potential demand for TVET is around 265,000, including school dropouts at GCE AL, GCE OL, and below GCE OL.

Both public and non-public TVET service providers offer more than 450 programs per year covering 23 different fields. Among them, nine fields account for more than 70% of enrollment and graduate output of the TVET sector. Of these, ICT, building and construction, and community development fields account for 48% of total enrollment and 50% of graduate output. By gender, female enrollment in TVET is 41%, and it is over 70% in the textiles and garments, medical and health science, and office management fields. In terms of market demand, however, major reforms are needed to align the program mix with emerging demand patterns of the labor market. The National Vocational Qualification Framework of Sri Lanka plays a vital role in improving the quality and relevance of TVET programs and creating career development pathways for TVET graduates. It has made significant progress since its introduction in 2004, and the number of certificates issued increased by 38% from 47,107 in 2016 to 65,212 in 2017. About 50% of National Vocational Qualification (NVQ) certificates issued in 2017 were at NVQ Level 3 and another 25% were at NVQ Level 2 or below.[7] In fact, from 2016 to 2017, there was a significant increase in the output of NVQ Level 1 and Level 2 certificates, but a marginal drop in NVQ Level 4 certificates. Public sector contributions to NVQ certification in 2016 was 74% and private sector contributions, 26%. Key challenges of the NVQ system are long delays in getting the courses accredited and satisfying quality assurance requirements in training. Sri Lanka's school education and TVET sectors are not strongly geared to supply the core competencies and technical skills demanded by employers. As a result, the skill sets of most school leavers are below the standards that employers expect in the workers they would like to hire. A review of the underlying evidence is in the following sections.

---

[3]   In Table 2.1, GCE OL refers to students who have passed the GCE OL examination, which is taken in Grade 11. Students who have not passed the GCE OL examination are not considered as having attained OL and are categorized with Grades 6–10.

[4]   Compulsory subjects: Own language, English, Mathematics, Science, History, and Religion. Optional subjects: Civics, Arts, Dancing, Commerce, Entrepreneurship, and Agriculture.

[5]   Arts: Includes many social science and humanities subjects. Science: There are two streams for science students–bioscience and physical science. Commerce: Accounts, Commerce, Economics, Entrepreneurship, etc.

[6]   Private and nongovernment organization (NGO) service providers can be further classified under five categories: (i) for-profit providers of institution-based individual training, (ii) NGO providers, (iii) fee-based training provided by companies in the fields of expertise, (iv) professional institutions, and (v) Chambers of Commerce and Industry. For further details, see Dundar et al. 2014.

[7]   NVQ level 1 is a National Certificate course; levels 2–4 are National Certificate courses; levels 5–6 are National Diplomas; and level 7 is equivalent to a bachelor's degree.

General education outcomes

The National Education Research and Evaluation Centre periodically evaluates the performance of the general education system by assessing the learning outcomes of secondary school students when they are between two milestone examinations: the fifth-year scholarship exam and the GCE OL. The most recent performance evaluation of students in Grade 8 in science and mathematics shows that, in 2016, 50% of Grade 8 students scored less than 47.5 (out of 100) in mathematics while 66% scored less than 50 in science. Slightly more than two-thirds (68%) scored less than 40 (out of 100) in English (NEREC 2017a). There was hardly any change in performance in these three subjects between 2014 and 2016 (Table 2.2). A comparison of competency levels in the use of visual clues and contextual clues in vocabulary and the mechanics of writing suggests a marked decline in achievements during 2014–2016 (Table 2.3).

Table 2.2: Learning Achievements of Grade 8 Students, 2014 and 2016

Subject	2014		2016	
	Mean	SD	Mean	SD
Mathematics	50.87	20.29	51.11	20.23
Science	41.16	20.92	41.76	20.73
English language	35.23	18.32	35.81	18.93

SD = standard deviation.
Note: Mean based on scores out of 100.
Source: National Education Research and Evaluation Centre (2017a).

Table 2.3: Comparison of Competency Levels Related to the English Language, 2014 and 2016

Competency	Competency Level	2014 (%)	2016 (%)	Percentage Point Change
Vocabulary	Uses English words in the proper contexts	49.6	56.7	7.1
	Uses the dictionary effectively	40.4	42.7	2.3
	Uses visual clues and contextual clues to derive the meaning of words	54.3	47.8	(6.5)
Reading	Transfers information into other forms	35.7	36.4	0.7
	Extracts the general idea of a text	46.5	49.2	2.7
Grammar	Analyze the grammatical relations within a sentence	45.4	44.1	(1.3)
	Construct complex sentences through the process of subordination	44.4	48.9	4.5
Mechanics of Writing	Uses commas with understanding	47.3	34.1	(13.2)

( ) = negative.
Source: National Education Research and Evaluation Centre (2017a).

Learning outcomes, as measured by internationally comparable indicators in the Trends in International Mathematics and Science Study (TIMSS), are not encouraging either. The TIMSS found a national average score of 23.16 with a standard deviation of 13.51 (Table 2.4). The mean value for rural students is 21.77 and for boys, 22.19. In fact, from 2014 to 2016, overall performance declined while there was little satisfactory progress in students' achievements in cognitive and content domains assessed using 15 standard questions covering three criteria (knowing, applying, and reasoning). Students managed to reach 50 or above in arithmetic only for 2 out of 15 questions. In algebra, students achieved 50 or more only in 4 out of 15 questions in both years (Table 2.5). Outcomes in geometry were discouraging with none of the questions scoring more than 50% in either 2014 or 2016. (Table 2.6). Skills in data and chance scored 50% and above only in three items in 2014, and in four items in 2016 (NEREC 2017b).

Table 2.4: Learning Achievements in TIMSS, 2014 and 2016

	2014		2016	
	Mean	SD	Mean	SD
National	23.31	13.60	23.16	13.51
Rural	21.00	11.46	21.77	12.29
Urban	27.89	16.11	26.78	15.71
Female	24.34	13.67	24.05	13.25
Male	22.22	13.43	22.19	13.72

SD = standard deviation, TIMMS = Trends in International Mathematics and Science Study.
Note: Mean based on scores out of 100.
Source: National Education Research and Evaluation Centre (2017b).

Table 2.5: Comparison of Students' Achievement in Relation to the Content Domain – Algebra

Domain	Question	Question Number	2014 (%)	2016 (%)
Knowing	Algebraic terms	1	58.3	57.8
	Algebraic expressions	5	23.2	24.0
	Algebraic expressions	9	26.6	26.9
	Algebraic expressions	32	27.2	25.2
	Algebraic expressions with brackets	42	3.2	7.8
Applying	Equations and formulas	14	52.2	50.6
	Number patterns	18	50.9	51.6
	Algebraic expressions	22	29.9	30.9
	Equations and formulas	36	10.6	24.6
	Equations and formulas	40	2.0	3.6
Reasoning	Algebraic expressions	25	32.7	34.4
	Number patterns	26	51.2	53.1
	Algebraic expressions by substituting integers	28	17.9	24.3
	Number patterns	45	21.3	23.1
	Algebraic expressions	48	0.7	0.4

Source: National Education Research and Evaluation Centre (2017b).

**Table 2.6: Comparison of Students' Achievement in Relation
to the Content Domain – Geometry**

Domain	Question	Question Number	2014 (%)	2016 (%)
Knowing	Locations and spatial relationships	8	29.0	28.5
	Measurement (units of time)	13	37.0	35.2
	Drawing plane figures to scale	30	0.5	1.7
Applying	Knowledge of cube nets	4	33.3	34.3
	Measurement(length)	17	27.8	27.6
	Measurement (time)	34	9.7	10.6
	Measurement (area of a triangle)	38	25.0	22.2
Reasoning	Values of angles	21	20.7	19.8
	Ratios in terms of fractions	24	21.4	21.5
	Measurement (area of a rectangle)	41	15.9	17.1

Source: National Education Research and Evaluation Centre (2017b).

Technical and vocational education and training sector

The TVET sector draws its trainees from the general education system and, potentially, the 275,000 young people who enter the job market can be regarded as the sector's target clientele (Appendix, Figure A.1). There are four major types of new entrants to the training sector: (i) school dropouts before GCE OL (16.4%); (ii) after GCE OL (26.9%); (iii) after GCE AL (28.3%); and (iv) candidates with higher education qualifications, e.g., AL plus (24.7%) [8] The estimated total enrollment in public, private, and nongovernment organization (NGO) sector institutions is around 200,000 per year, and the number of individuals without any training opportunity is around 65,000 per year. The public sector accounts for about 77% of TVET enrollment, while the rest is shared by private and NGO service providers (TVEC 2017b). This may go up to 140,000 if average annual TVET dropouts numbering about 80,000 students are included.[9] The high dropout rates have been attributed to a range of reasons, including poverty, poor quality of programs, inadequate OJT, long duration of programs, difficulties in following theory course modules, and lack of employment opportunities. The total number of unemployed without vocational training, however, was around 260,000 in 2016, and this may be due to the cumulative effect of the unemployed without vocational training.

---

[8]   The rest joins the labor market without any training.

[9]   It should be noted that the newly introduced policy of 13 years of guaranteed education is still being piloted and there may be a lag before impact in data can be seen. The new policy comprises three stages from October 2017 to 2020.

Notwithstanding the large number of young Sri Lankans requiring job-oriented training, the system's capacity to generate the required training is constrained. In terms of learning outcomes, the TVET sector performs no better than the general education system. A sizable proportion of TVET graduates leave training programs without the skills that employers require. Tracer studies on the employability of TVET graduates reveal a high rate of unemployment among TVET graduates who had been trained for employment in the ICT, construction, tourism, and light engineering sectors (ADB 2018). For example, the rate of job placement was lowest among TVET ICT graduates compared with graduates of metal and light engineering, hotel and tourism, construction, and other sector-specific programs (Table 2.7). The rate of job placement was significantly lower relative to graduates with NVQ Certificate and NVQ Degree level qualifications (Table 2.8). While, overall, 28% of all unemployed individuals had TVET qualifications, by gender, it was 23% among males and 32% among females (TVEC 2017a).

Table 2.7: Job Placement Rate of TVET Graduates by Sector, 2016

Sector	Male	Female
Construction	66.3	54.6
Hotels and Tourism	73.2	–
ICT	50.9	38.0
Metal and Light Engineering	73.5	–
Others	56.2	40.2

– = none, ICT = information and communication technology, TVET = technical and vocational education and training.
Source: ADB (2018).

Table 2.8: Job Placement Rate of TVET Graduates by Qualifications, 2016

Sector	Male	Female
Non-NVQ	61.0	36.7
NVQ Certificate	66.2	40.2
NVQ Degree	84.0	77.8

NVQ = National Vocational Qualification, TVET = technical and vocational education and training.
Source: ADB (2018).

There seems to have been little change in the share of unemployed among those with and without TVET since the 1990s (Figure 2.12). Of those unemployed without TVET, 40% did not have GCE OL qualifications in 2017 (Figure 2.13). The share of females among those without TVET is much higher at around 55%, with 41% of those unemployed females without TVET are at least AL qualified (Table 2.9). Recent evidence on the labor market readiness of first-time job seekers suggests that about 40% of TVET graduates with lower than secondary-level education are not ready for paid work. Similarly, about 32% of new entrants to the labor market with only a university degree/higher education institute, and 33% of those with TVET plus secondary education, are insufficiently trained for work. However, individuals with a university degree or higher education qualifications plus TVET qualifications were rated suitable for immediate employment by the highest proportion (69%) of employers (Table 2.10).

Figure 2.12: Vocational Training and Unemployment, 1998–2017

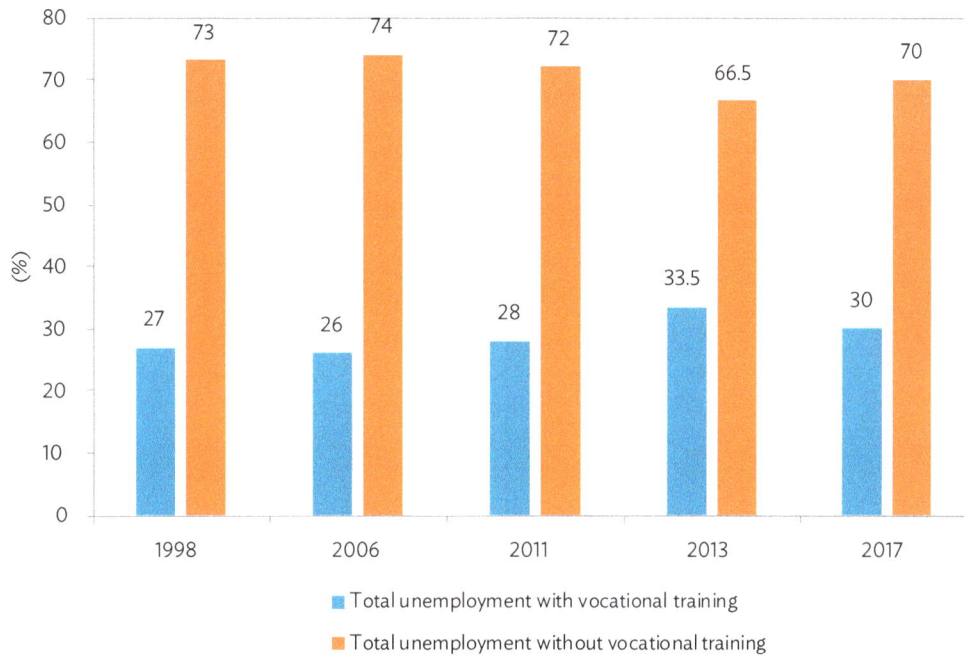

Source: Tertiary and Vocational Education Commission. Labour Market Information Bulletin. Colombo (Years: 1998, 2002, 2011, 2013, 2017).

Figure 2.13: Unemployed Persons without Vocational Training by Level of Education

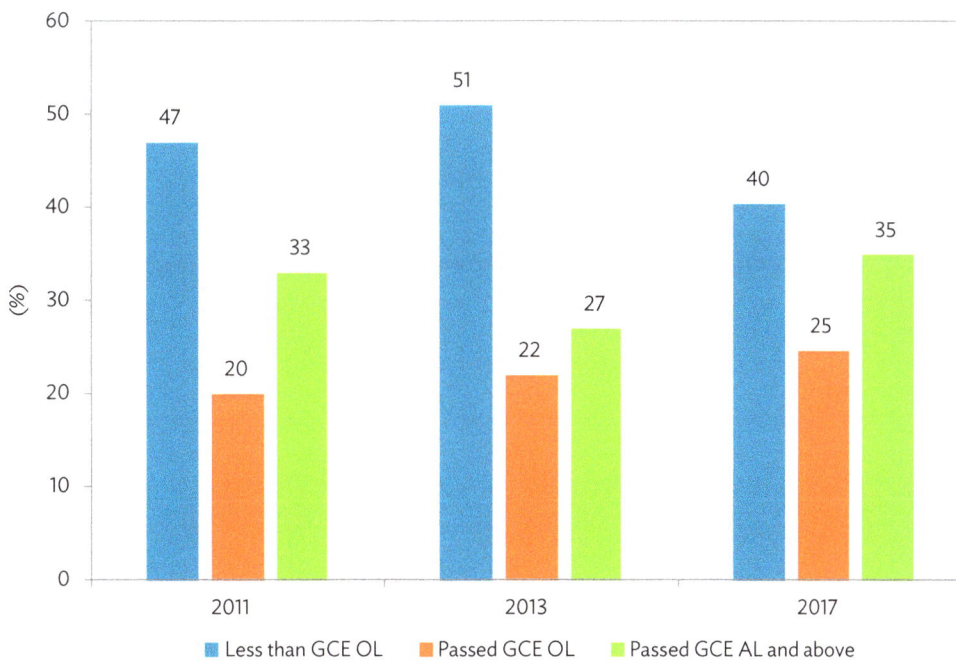

GCE AL = General Certificate of Education advanced level, GCE OL = General Certificate of Education ordinary level.
Sources: Tertiary and Vocational Education Commission. Labour Market Information Bulletin. Colombo (Years: 2012, 2014, and 2017).

Table 2.9: Unemployed Persons without TVET by Gender and Level of Education, 2016

Level of Education	Male		Female		Total	
	%	Number	%	Number	Number	%
Less than GCE OL	56	65,827	39	55,717	121,544	47
GCE OL	21	24,360	20	28,180	52,540	20
GCE AL and above	24	28,220	41	59,141	87,360	33
	100	118,407	100	143,038	261,444	100

GCE AL = General Certificate of Education advanced level, GCE OL = General Certificate of Education ordinary level, TVET = technical and vocational education and training.
Source: TVEC (2017a).

Table 2.10: Preparedness for Work of First-Time Job Seekers by Educational/Training Attainment, 2017

Academic Qualifications	Preparedness (%)			
	Well	Average	Poor	Total
Only secondary education	31	50	19	100
TVET and lower than secondary education	23	38	39	100
Only university or HEI	32	36	32	100
Secondary school and TVET qualifications	27	40	33	100
University / HEI with TVET qualifications	69	26	5	100

HEI = higher education institute, TVET = technical and vocational education and training.
Source: Department of Census and Statistics (2017b).

TVET programs offered lack labor market relevance and the skill mix of TVET graduates does not match market demand. Employers in emerging sectors look for workers with high cognitive, noncognitive, and technical skills, and not just routine cognitive and manual skills.[10] The analysis of returns to training also suggest that it is competencies rather than qualifications that offer higher returns (Dundar et al. 2017). Modern workplace skill requirements include communication, leadership, organizing, and innovative skills, but Sri Lanka's educational and training institutions is not producing enough workers with market-relevant technical and managerial skills. There are three key reasons for this skills mismatch: (i) poor alignment of curricula with competencies required to qualify for specific occupations; (ii) a disproportionate number of students pursuing trades for which there is little demand; and (iii) a training system unable to adjust to changes in the demand for skills fueled by growth and technological change (Dundar et al. 2017). The participation of the private sector in curriculum development providing internships and training for students and teachers is insufficient.

---

[10]   For details, see Dundar et al. (2017) and Dundar et al. (2014).

2.3   Selection of Subsectors for Analysis

Selection of the two subsectors for skills gap analysis was based on several criteria. First, the list of 30 priority subsectors with export growth potential was obtained from the Board of Investment's (BOI) study, *Targeting Sectors for Investment and Export Promotion* (Board of Investment of Sri Lanka, Export Development Board of Sri Lanka, and Harvard Center for International Development 2017). These priority sectors were identified by constructing three composite indices, each made up of several indicators. The three composite indices were market opportunities (domestic, regional, and world market opportunities); investor interest (investor interest in Sri Lanka, investors' access to regional markets and peer country exports); and the impact on Sri Lanka's economy (in terms of job creation potential, job quality potential, potential beyond the Western Province, and the potential for new linkages). Second, in addition to the criteria used by the BOI, several other criteria are considered, including the relative magnitude of employment in the BOI-selected priority sectors, the share of women in the workforce employed in each sector, the degree of export orientation, attractiveness to foreign direct investment, technology orientation, growth potential, and priority rankings by the Export Development Board (EDB). The key priority sectors identified earlier by policy makers (construction, ICT, tourism, and light engineering) were not considered as these four sectors have been the subject of previous analyses of skills gaps (see Chandrasiri and Gunatilaka 2015, 2016; Chandrasiri, De Mel, and Jayathunge 2017). Table 2.11 sets out information about each of the BOI's priority sectors related to these criteria.

Of the 30 sectors listed in the table, the electronics and electrical equipment is covered under the electronics and electrical equipment (EE) subsector while food products and beverages is included in the food and beverages (FB) subsector. These sectors emerged as key economic sectors for a detailed analysis on skills gaps out of a total of nine major sectors identified by the EDB as shown in last column of Table 2.11.[11] Table 2.12 also includes two additional sectors for analysis, namely health and care work, and hospitality as these two have a large share of female employment. Ultimately, the EE and FB were chosen for this study as these two subsectors have been identified as priority areas by recent studies. For example, according to the Board of Investment of Sri Lanka, Export Development Board of Sri Lanka, and Harvard Center for International Development (2017) study, the electronic sector ranked first and the electrical sector ranked fifth, while food products ranked ninth and the beverages sector ranked 29th. Similarly, the list of priority sectors identified by the National Export Strategy (NES) 2018–2022 includes FB and EE machinery as high-priority sectors for export development (EDB 2018a).

The NES acknowledges inadequate supply of semiskilled and skilled workers as a constraint on export growth of the EE subsector. To address this issue, the NES has also formulated five action programs that focus on stakeholders on the supply side of the labor market.

---

[11]   In this analysis, the FB subsector and the EE subsector are defined in terms of five-digit International Standard Industrial Classification (ISIC) subsector codes as follows: FB: ISIC subsectors 10101 to 10802 and subsectors 11011 to 11049; EE: ISIC subsectors 26100 to 26409 and ISIC subsectors 27101 to 27900.

Table 2.11: Criteria for Selecting Priority Subsectors

Subsectors	Exports 2017 ($ million)	Imports 2017 ($ million)	Employment % (%)	Labor Productivity 2017 (SLRs million)	Value of Exports 2017 ($ million)	BOI Firms (no.)	Growth of Exports, 2007-2016 (% per year)	BOI Prioritization: Priority Score	BOI Prioritization: Potential Economic Impact	BOI Prioritization: Market Opportunity and Investor Interest	BOI-Overall Priority Rating: BOI	EDB: NES Sectors
Electronics	149	779.2	0.04	0.84	99	12	(5.36)	High	High	High	Highest	Yes
Industrial machinery and equipment	203.3	1,141.5	0.01	1.74		19	20	High	High	High	Highest	Yes
Transport equipment (motor vehicles and trailers)	40.4	1,250.7	0.02	2.30	39	5	(2.72)	High	High	High	Highest	
Accommodation and food services (tourism)	2,431.1	1,262.5	1.97	1.32	3,925	249		High	Average	High	Highest	Yes
Electrical equipment	240.5	497.9	0.15	2.50	59	23	(18.36)	High	High	High	Highest	Yes
Fabricated metal products	38.4	476.5	0.24	1.03	116	36	5.25	High	High	High	Highest	
Chemical products	225.7	1,849	0.55	1.63	146	33	7.15	High	Average	High	Highest	
Other products (medical, decorative, recreational)	301.2	244.9	0.18	0.79		132	0.25	Average	Average	Average	Highest	
Food products	1,204.4	1,505	3.53	1.11		93		Average	Low	High	Highest	Yes
Business administration and support (BPO)	42.6	58.3	0.63	...	226	72	2.01	Average	Average	High	Highest	Yes
Basic metals	99.7	902.5	0.02	3.27		17	15.47	Average	Average	Average	High	
Software and IT	627.9	300.3	0.95	0.25	724	108	(2.73)	Average	Average	Average	High	Yes
Cement, ceramics, glass, and other metal products	97.5	605.4	0.43	1.34	34	50		Average	High	Average	High	
Transportation and storage (logistics)	1,923.2	1,461.6	6.49	2.86		82		Average	Average	Average	High	
Finance and insurance	370.9	439.1	2.00	4.04		41		Average	Average	Average	High	

continued on next page

Table 2.11 continued

Subsectors	Exports 2017 ($ million)	Imports 2017 ($ million)	Employment % (%)	Labor Productivity 2017 (SLRs million)	Value of Exports 2017 ($ million)	BOI Firms (no.)	Growth of Exports, 2007–2016 (% per year)	BOI Prioritization Priority Score	Potential Economic Impact	Market Opportunity and Investor Interest	BOI-Overall Priority Rating BOI	EDB NES Sectors
Rubber and plastic products	1,043.9	445.6	0.51	1.97	910	104	10.36	Average	Average	Average	High	
Pharmaceutical products	8.4	398.2	0.04	2.03		8		Average	High	Average	High	
Refined petroleum and coke products	86.6	2,854.8	0.04	62.17	434	1	56.3	Average	Average	Average	High	
Education	0.0	0.0	4.33	0.77		16		Average	Average	Low	High	
Transport equipment (ships, motorcycles, others)	91.0	2,885.2	0.00	...	124	12	(9.64)	Average	Average	Average	High	Yes
Wearing apparel	5,271.2	844.1	6.48	0.75	4,739	277	(5.20)	Average	Average	Average	Lower	
Footwear, leather, travel goods, and related products	63.2	60.3	0.12	0.75	158	11	(5.10)	Average	Average	Average	Lower	
Paper products	48.2	494	0.09	2.12	69.5	22	(8.60)	Average	Average	Average	Lower	
Furniture	29.5	37.7	0.40	0.35		10		Average	Average	Average	Lower	
Textiles	355.1	1,391.5	0.70	0.88	293	36	5.28	Average	Average	Average	Lower	
Construction	58.1	28.8	5.79	2.03		134		Average	Average	Average	Lower	
Products of wood, cork and straw except furniture	46.0	76.0	0.53	0.21	69.5	10	(2.20)	Low	Average	Low	Lower	
Printing and recorded media	3.2	5.5	0.37	1.30		3		Low	Average	Low	Lower	
Beverages	27.4	61.8	0.10	2.61		33	(23.00)	Low	Low	Low	Lower	Yes
Tobacco	56.0	8.3	0.26	3.19		2	2.4	Very Low	Very Low	Very Low	Lower	

... = not available, ( ) = negative, BOI = Board of Investment, BPO = Business Process Outsourcing, EDE = Export Development Board, IT = information technology, NES = National Export Strategy.
Sources: Department of Census and Statistics (2017b); Board of Investment, Export Development Board of Sri Lanka, and Harvard Center for International Development (2017); and Export Development Board (2018a).

**Table 2.12: Criteria Related to the Food and Beverages,
and Electronics and Electricals Subsectors**

Criteria	Unit	Food and Beverages	Electronics and Electrical Equipment	Hospitality	Health and Care Work
Relative share of total employment (LFS 2016)	%	3.5	0.2	2.6	1.8
Relative share of female employment (LFS 2016)	%	4.7	0.2	2.6	3.1
Share of female employment within subsector (LFS 2016)	%	45.6	44.5	35.0	62.0
Growth of employment per year, 2012-2016	%	30.9	...	10.0	20.0
Share of employment (LDS 2017)	%	5.1	0.3	4.2	4.0
Share of vacancies (LDS 2017)	%	11.3	14.3	8.6	8.2
Share of total exports	%	8.1	2.6	16.0	...
Share of BOI firms	%	7.6	2.1	15.1	...
BOI priority status		Average	High	High	...
NES priority status		High	High	High	...

... = not available, BOI = Board of Investment, LDS = Labour Demand Survey, LFS = Labour Force Survey, NES = National Export Strategy.
Sources: Employment-related statistics estimated with microdata from the Department of Census and Statistics Quarterly Labour Force Surveys of 2012 and 2016, and Labour Demand Survey of 2017. Other indicators from Board of Investment, Export Development Board of Sri Lanka, and Harvard Center for International Development (2017); and Export Development Board (2018a).

In terms of technology orientation, the FB subsector is a resource-intensive industry while the EE subsector is classified as a differentiated technology subsector. Both subsectors have long been identified as potential subsectors for growth and export promotion in Sri Lanka. For example, Lall et al. (1996) identified food products and electronics as a potential subsector for export growth.[12] Similarly, from 1994 onward, several studies and development plans of various policy regimes have recognized FB as a subsector with high capacity to absorb local resources including labor and the potential to produce value-added products for export markets. However, full growth potential of these subsectors in terms of exports and employment have not been realized, owing partly to limited implementation of policy recommendations and developments related to these subsectors.

The FB subsector is an important subsector in manufacturing which has benefited from both open and close economy policy regimes since the early 1970s. As a resource-based industry, the FB subsector is less oriented toward research and development-based, new product development and exports. Within the FB subsector, key production activities include (i) processing and preserving of meat, fish, crustaceans and mollusks, fruits, and vegetables; (ii) manufacture of vegetable and animal oils and fats, dairy products, grain mill products, starches and starch products, bakery products, sugar, cocoa, chocolate and sugar confectionery, macaroni, noodles, couscous and similar farinaceous products; (iii) manufacture of prepared animal feeds; and (iv) manufacture of soft drinks, and production of mineral waters and other bottled waters. The estimated number of industrial establishments in the FB subsector is around 5,073 (DCS 2019). Its export intensity per employee is $1.40 as against $22.10 per employee in the EE subsector. The EE subsector is relatively new in manufacturing. It began as a service-oriented sector and gradually transformed into a manufacturing industry. This transformation is directly linked with

---

[12] Lall et al. (1996) also identified six industry clusters as a strategy to promote manufactured export growth: (i) food products, (ii) rubber products, (iii) textiles and garments, (iv) gems and jewelry, (v) leather and footwear, and (vi) electronics and software.

global value chains to which Sri Lanka is relatively a new entrant. Its competitiveness in the global market would depend, among other factors, on availability of labor with necessary skills. There are an estimated 264 industrial establishments in the EE subsector involved primarily in the manufacture of (i) electronic components and boards; (ii) electric motors, generators, transformers, and electricity distribution and control apparatus; (iii) communication products; (iv) optical instruments and photographic equipment; (v) other electronic and electric wire and cables; (vi) wiring devices; (vii) batteries and accumulators; (viii) electric lighting equipment; and (ix) other electrical equipment.

According to data from the Labour Force Survey 2016, the FB subsector employed 276,707 workers in 2016. The EE subsector, in contrast, was much smaller, employing 12,708 workers. Table 2.13 sets out the estimates of employment in the two subsectors according to three sources of sample survey data: the Labour Force Survey, which is a household survey; and the Labour Demand Survey 2017 and the Annual Survey of Industries 2016, which are both surveys of establishments. To analyze the distribution of employment in the two subsectors by occupation, education, and training in section 2.4, the Labour Force Survey data is used as it is the only source of information on the characteristics of the workforce employed in the subsectors.

Figure 2.14: Export Performance of the FB and EE Subsectors, 2010–2018

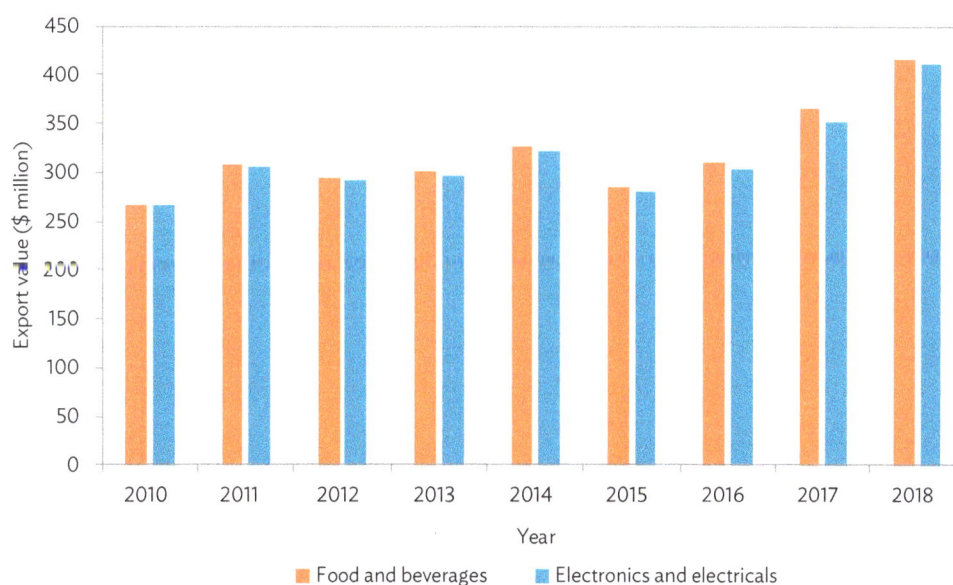

EE = electronics and electricals, FB = food and beverages.
Source: Export Development Board (2018a).

Table 2.13: Employment in the FB and EE Subsectors according to Different Data Sources

Survey	Total Employment	
	FB	EE
Labour Force Survey 2016	276,707	12,708
Labour Demand Survey 2017	261,424	15,947
Annual Survey of Industries 2016	282,843	14,981

EE = electronics and electricals, FB = food and beverages.
Source: Authors' compilation from the Department of Census and Statistics surveys mentioned in the table.

2.4   Skills-Related Characteristics of the Food and Beverages, and Electronics and Electricals Subsectors

The FB subsector is at least 16 times as large as the EE subsector in terms of total employment. However, the EE subsector employs a far larger share of workers in the higher-level occupations of managers, professionals, and technicians (36%) than the FB subsector (11%) (Table 2.14). The shares of craft workers and elementary workers in the FB subsector are also much higher, whereas in the EE subsector, machine operators and technicians predominate. Women account for nearly half of employment in the FB subsector and only a little more than a third in the EE subsector. And, in both subsectors, women have higher employment shares in the lower-skilled occupations, whereas men's employment shares are much higher in the higher-skilled occupations.

Figure 2.15 presents the age profiles of the workforce in the two subsectors and they provide very interesting contrasts. Workers in the FB subsector tend to be older, with those between 30 and 60 years of age accounting for a little more than two-thirds of the workforce, while those between 20 and 24 years of age account for only 7%. In contrast, nearly 30% of workers in the EE subsector are between 20 and 24 years of age, signaling that the subsector itself developed relatively recently, and that young people may find this subsector relatively attractive.

The two subsectors also differ in terms of organizational structure: the much smaller EE subsector displays the characteristics of more formal industrial organizations (Figure 2.16). Nearly half of all those employed in the FB subsector are not attached to any specific institution, but 18% are attached to firms with at least 100 workers. In contrast, nearly 60% of those employed in the EE subsector are in firms that employ at least 100 workers. Around 10% of workers in the FB subsector work in microenterprises, and a slightly smaller proportion of workers in the EE subsector are employed in firms with 5–9 workers. Workers in the EE subsector appear better educated than those in the FB subsector (Figure 2.17). In the EE subsector, all workers are primary educated or more, with 10% of workers having degrees (as compared with 1.4% in the FB subsector). At the other end of the education scale, 30% of EE workers are secondary educated compared with 53% of FB workers. The status of training in the two subsectors is also very low; only 6% of those employed in the FB subsector have some formal training, whereas this ratio is 25% in the EE subsector (Table 2.15). The low level of education and inadequate training among workers in the two subsectors highlight the need for skills development both before and after employment in the two subsectors. Section 4.5 provides a more detailed discussion on the type of skills development needed.

Table 2.14: Employment in the FB and EE Subsectors by Occupation and Gender, 2016

	Food and Beverages			Electronics and Electricals		
	All	**Men (53%)**	**Women (47%)**	**All**	**Men (65%)**	**Women (35%)**
Managers	5.4	9.0	1.3	12.6	19.3	0.0
Professionals	1.0	1.1	1.0	1.4	0.0	4.1
Technicians	4.5	7.1	1.6	13.2	17.4	5.2
Clerical workers	3.1	3.2	2.9	20.4	12.8	34.7
Service workers	7.2	6.0	8.7	2.5	3.8	0.0
Skilled agricultural workers	2.7	1.4	4.0	0.0	0.0	0.0
Craft workers	35.3	28.1	43.4	18.7	14.7	26.4
Machine operators	12.0	16.9	6.5	29.8	32.0	25.5
Elementary workers	28.8	27.3	30.5	1.4	0.0	4.1
Total	**100.0**	**100.0**	**100.0**	**100.0**	**100.0**	**100.0**
Total (number)	**276,707**	**146,703**	**130,004**	**12,708**	**8,311**	**4,397**

EE = electronics and electricals, FB = food and beverages.
Source: Authors' estimates based on microdata from the Department of Census and Statistics Quarterly Labour Force Survey 2016 using sample weights.

Figure 2.15: Age Distribution of the Workforce in the FB and EE Subsectors (%), 2016

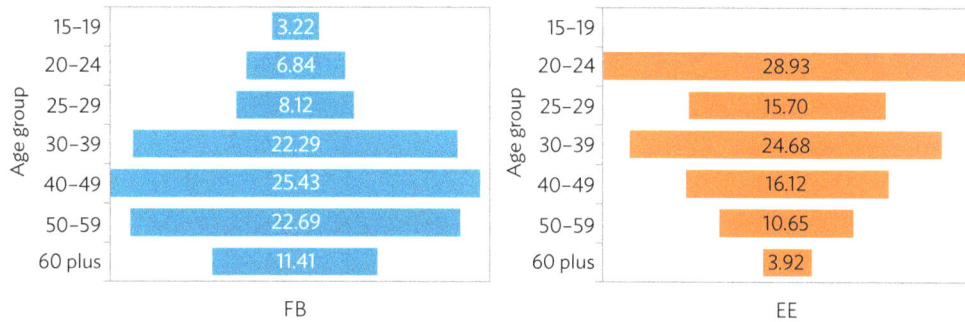

FB

Age group	%
15–19	3.22
20–24	6.84
25–29	8.12
30–39	22.29
40–49	25.43
50–59	22.69
60 plus	11.41

EE

Age group	%
15–19	
20–24	28.93
25–29	15.70
30–39	24.68
40–49	16.12
50–59	10.65
60 plus	3.92

EE = electronics and electricals, FB = food and beverages.
Source: Authors' estimates based on microdata from the Department of Census and Statistics Quarterly Labour Force Survey 2016 using sample weights.

Figure 2.16: Distribution of Employment by Size of Firm in the FB and EE Subsectors, 2016

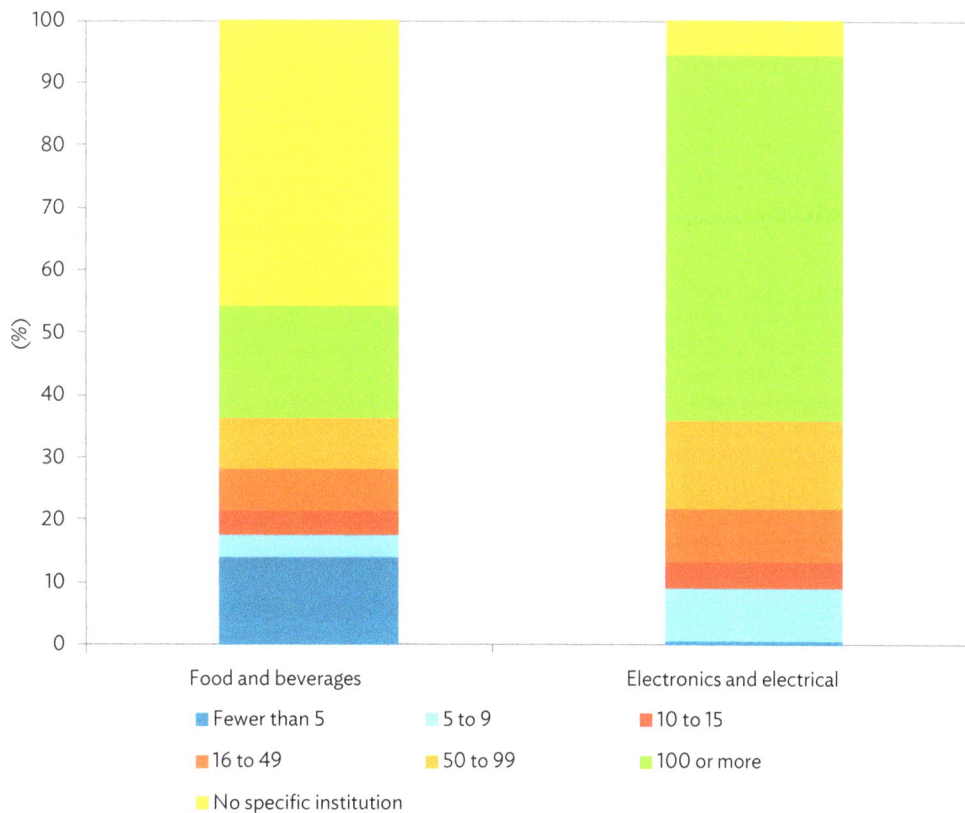

Food and beverages   Electronics and electrical

Fewer than 5 | 5 to 9 | 10 to 15 | 16 to 49 | 50 to 99 | 100 or more | No specific institution

EE = electronics and electricals, FB = food and beverages.
Source: Authors' estimates based on microdata from the Department of Census and Statistics Quarterly Labour Force Survey 2016 using sample weights.

Figure 2.17: Distribution of Employment by Occupation and Education in the FB and EE Subsectors, 2016

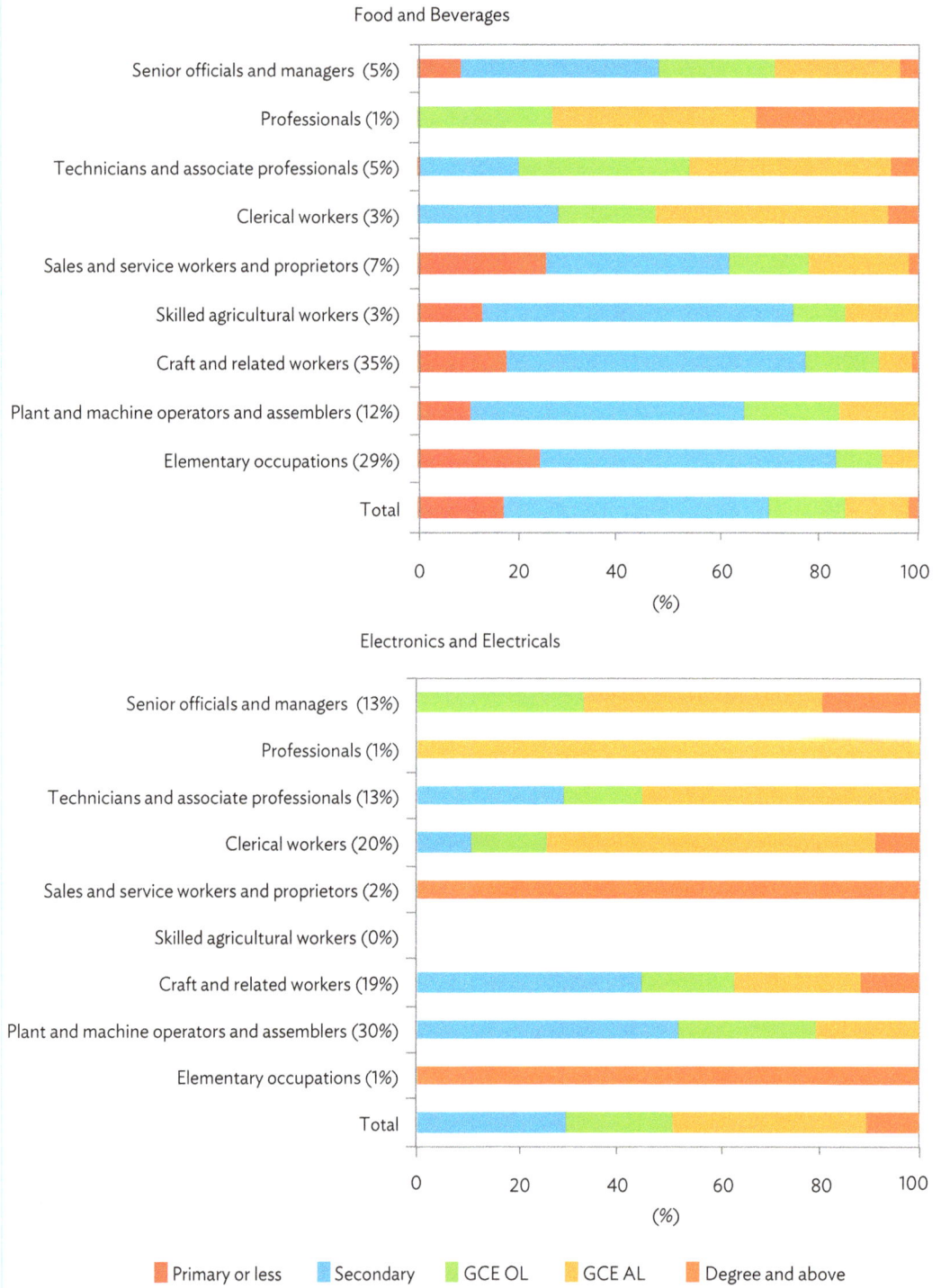

Food and Beverages

Electronics and Electricals

Legend: ■ Primary or less ■ Secondary ■ GCE OL ■ GCE AL ■ Degree and above

EE = electronics and electricals, FB = food and beverages, GCE AL = General Certificate of Education advanced level, GCE OL = General Certificate of Education ordinary level.
Source: Authors' estimates based on microdata from the Department of Census and Statistics Quarterly Labour Force Survey 2016 using sample weights.

Table 2.15: Distribution of Employment by Training in the FB and EE Subsectors (%), 2016

	No Training	General	Diploma	Higher Diploma	General and Diploma	Diploma and Higher Diploma
Food and Beverages						
Senior officials, managers, and legislators	88.1	9.8	2.1	0.0	0.0	0.0
Professionals	63.4	8.5	0.0	17.2	0.0	10.9
Technicians and associate professionals	63.3	21.9	5.3	6.9	2.6	0.0
Clerical workers	80.3	7.9	6.7	0.0	2.0	3.1
Sales and service workers and proprietors	92.2	5.6	2.2	0.0	0.0	0.0
Skilled agricultural workers	100.0	0.0	0.0	0.0	0.0	0.0
Craft and related workers	92.1	6.4	1.0	0.5	0.0	0.0
Plant and machine operators and assemblers	90.1	6.1	1.6	2.1	0.0	0.0
Elementary occupations	98.7	1.3	0.0	0.0	0.0	0.0
Total	**91.8**	**5.6**	**1.2**	**0.9**	**0.2**	**0.2**
Total (number)	254,072	155,57	3,441	2,569	496	573
Electronics and Electricals						
Senior officials, managers, and legislators	35.5	41.9	22.6	0.0	0.0	0.0
Professionals	0.0	100.0	0.0	0.0	0.0	0.0
Technicians and associate professionals	46.9	53.1	0.0	0.0	0.0	0.0
Clerical workers	79.5	11.2	9.3	0.0	0.0	0.0
Sales and service workers and proprietors	100.0	0.0	0.0	0.0	0.0	0.0
Skilled agricultural workers	0.0	0.0	0.0	0.0	0.0	0.0
Craft and related workers	90.4	9.6	0.0	0.0	0.0	0.0
Plant and machine operators and assemblers	77.0	23.0	0.0	0.0	0.0	0.0
Elementary occupations	100.0	0.0	0.0	0.0	0.0	0.0
Total	**70.7**	**24.6**	**4.7**	**0.0**	**0.0**	**0.0**
Total (number)	8,980	3,124	603	0	0	0

EE = electronics and electricals, FB = food and beverages.
Source: Authors' estimates based on microdata from the Department of Census and Statistics Quarterly Labour Force Survey 2016 using sample weights.

A comparison of average daily cash wages by occupation in the two subsectors in Figure 2.18 shows that workers in the more skilled occupations enjoy higher mean wages in the FB subsector than in the EE subsector. In contrast, the average wage differential favors workers in the lower-skilled occupations in the EE subsector than in the FB subsector.

Figure 2.18: Average Real Daily Cash Wages in the FB and EE Subsectors, 2016

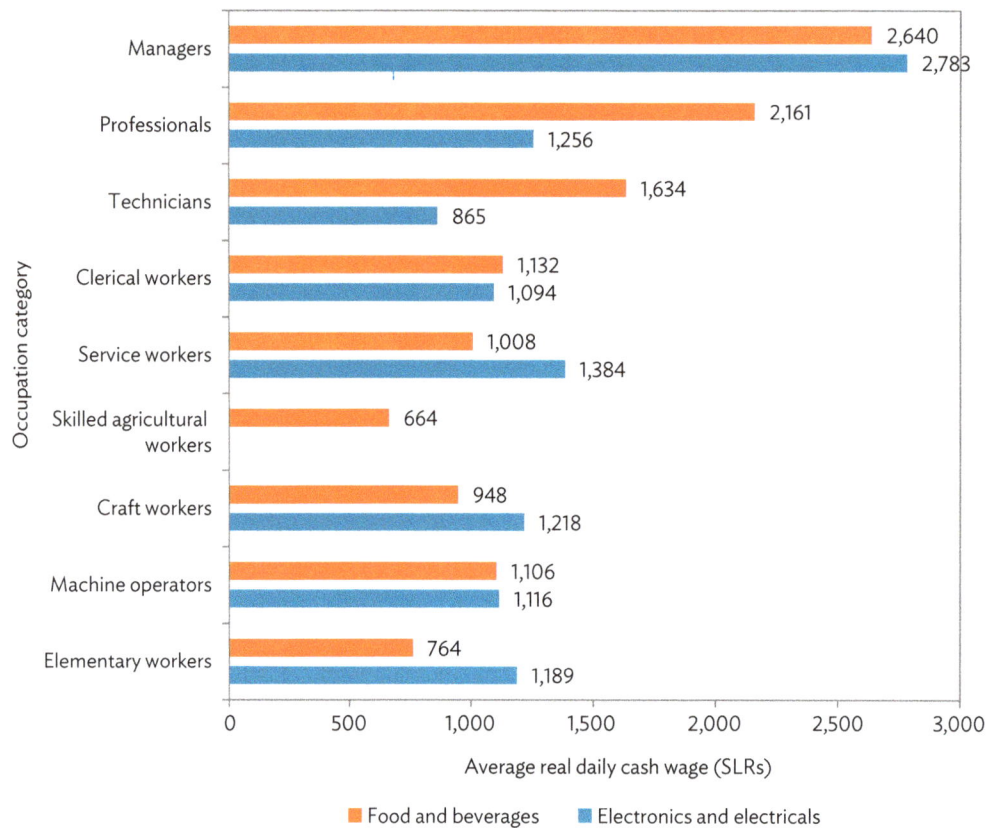

EE = electronics and electricals, FB = food and beverages.
Note: The Laspeyre's price index compiled by the Department of Census and Statistics, using Household Income and Expenditure Survey 2016, was used to adjust mean wage data for spatial differences in prices.
Source: Authors' estimates based on microdata from the Department of Census and Statistics Quarterly Labour Force Survey 2016 using sample weights.

CHAPTER 3
DEMAND FOR SKILLS

3.1   Data and Methodology

A rigorous estimation of skills deficits would ideally require longitudinal data about all the factors that have a likely impact on the demand for skills, including factors that impact on aggregate demand. Since Sri Lanka, like many other emerging economies, lacks such comprehensive data, the estimations in this study are based on available data and a series of strong assumptions.

This study looks at availability of skills in the two subsectors in three different dimensions: the demand for labor, qualitative skills deficits, and quantitative skills deficits. The demand for labor by occupation is estimated in quantitative terms using data on the growth performance and employment elasticity of the two subsectors from 2011 to 2015 (using output and employment data from the DCS's Annual Survey of Industries), and data about the occupation-wise distribution of employment are from the most recent establishment-level survey of labor demand, Labour Demand Survey 2017 (LDS 2017). The LDS 2017 does not contain information about skills deficits by occupation and industry sector, but provides the most recent data on total employment by sector and occupation as it is an establishment-level survey. To this estimated labor demand, the demand for foreign employment in the two subsectors, based on the estimates provided by the Sri Lanka Foreign Employment Bureau, are added.

The second dimension of the availability of skills is qualitative, that is, the availability of core competencies and technical skills. To estimate these skills deficits, the labor demand projections of the first dimension based on elasticities and applied occupation-wise skills-based weights derived from the Labour Force Survey of 2016 (LFS 2016) and the World Bank's STEP data of 2012 (STEP 2012) were applied to the projected employment totals. Information about education, training, and literacy in English was drawn from the LFS 2016, but since it is a household survey, it could not be regarded as being representative of industries. Since neither the LDS 2017 nor the LFS 2016 had information about core competencies and technical skills, STEP 2012 was used to extract information about the distribution of these skills. STEP 2012 is the only comprehensive source of skills availability data that is presently available.[13]

The methodology used for estimating qualitative skills is similar to that used in an earlier study (Chandrasiri and Gunatilaka 2016). That study looked at four different sectors and used STEP-derived skills deficits weights for the entire nonfarm sector to estimate and project skills deficits. It also assumed an annual employment growth rate of 5%. The methodology in this study is further refined. First, growth employment elasticities are used to estimate employment growth in each subsector as stated above. Second, STEP data were used to derive skills deficit weights by occupation, and for the two subsectors separately, instead of using nonfarm industries' wide weights as done earlier. This

---

[13]   The limitations of STEP 2012 are well known. It is a household survey rather than an establishments survey, so it is not representative of the economic structure of production units but of households. The STEP skills data set is based on self-assessment and subjective perceptions. While the data are not recent, in the absence of a suitable alternative, especially as the most recent LDS lacks information about skills by occupation level, STEP 2012 was used instead for the analysis.

chapter is concerned with these two dimensions of skills deficits only. The next chapter presents the third dimension, quantitative estimate of skills gaps that is based on training output of the technical and vocational education and training (TVET) and higher education sectors (using administrative data and labor demand estimates under the first dimension presented in this chapter). Chapter 4 also includes a brief note on the quality of school education to provide an overall understanding of supply-side issues constraining Sri Lanka's education and training systems.

Estimating labor demand using growth employment elasticities

The demand for labor has two components: domestic and foreign. Table 3.1 presents the projections of domestic demand based on employment elasticities and growth of output in the EE and FB subsectors from 2011 to 2015. Growth employment elasticity for the period 2011–2015 is estimated to be 1.8 for the FB subsector. In contrast, the EE subsector is relatively new and needs to attract export-oriented foreign direct investment through strong institutional and policy support. However, employment elasticity may not continue in the future, as GDP growth during 2011–2015 was above average and growth targets (e.g., 3.2% in 2019 to 5% by 2023) presented in the midterm macroeconomic framework (CBSL 2019) suggest the need for an adjustment in employment elasticity.

Two scenarios are projected based on both pessimistic and more optimistic assumptions. The major assumptions of the pessimistic employment projections include below 5% growth up to 2025, the absence of major structural reforms, and continued challenges facing institutional and policy support systems responsible for promoting export-oriented foreign direct investment. Accordingly, output growth of the EE subsector is assumed to be 9% per year over the next 5 years. It is also assumed that the employment elasticity of the EE subsector will reduce marginally after 2023 due to industry maturity effects (pessimistic scenario, Table 3.1). Similarly, output growth of the FB subsector is expected to be around 8% and employment elasticity, 0.8, over the next 3 years (2020–2022), and a marginal decline in both food and beverages due to industry maturity effects is assumed for the post-2022 period. The main assumptions of the optimistic employment projections include timely implementation of the National Export Strategy (NES) and New Trade Policy, active promotion of export-oriented foreign direct investment, and improvement in the overall political and business climate. In line with these considerations, it is assumed that both the EE and FB subsectors will continue to experience output growth with a marginal increase in employment elasticity over the next 5 years (optimistic scenario, Table 3.1).

Table 3.2 shows projected employment in the two subsectors in the two scenarios. According to LDS 2017 weights, the distribution of employment would be 22% (high-skill category), 7% (medium skill), and 71% (low skill) in the EE subsector, and 9% (high-skill category), 17% (medium skill), and 74% (low skill) in the FB subsector.

Demand projections for the EE and FB subsectors also include foreign demand for which official data released by the Foreign Employment Bureau are used.[14] Accordingly, foreign demand for the EE subsector occupations is 59 high-skilled and 96 medium-skilled jobs per year and that for the FB subsector occupations, 20 jobs per year. Foreign demand is highest for food technologists in the FB subsector. In the EE subsector, foreign employers are looking to hire workers for five major occupations: engineer – electrical, technician – electromechanical, electrical inspector, technician – electronics and telecom, and technician – industrial electrical. Foreign demand for FB and EE subsector workers has been projected assuming an annual growth rate of 3%.

---

[14]   These figures may not capture actual foreign demand for EE- and FB-related occupations as most of the medium- and high-skilled migrant workers find foreign employment through private sources.

Table 3.1: Growth Employment Elasticities Actual and Projected, 2011–2025

	Employment 2017	Estimates for 2011–2015		Adjusted 2017–2019		Adjusted 2020–2022		Adjusted 2023–2025	
		Employment elasticity	Output growth (%)	Employment elasticity	Output growth (%)	Employment elasticity	Output growth (%)	Employment elasticity	Output growth (%)
Pessimistic scenario									
EE	15,947	0.8	21.7	0.8	10	0.7	9	0.6	9
FB	261,424	1.8	18.1	1	10	0.8	8	0.7	7
Optimistic scenario									
EE	15,947	0.8	21.7	0.8	10	0.8	10	0.9	11
FB	261,424	1.8	18.1	1	10	0.8	11	0.9	12

EE = electronics and electricals, FB = food and beverages.
Sources: Employment data for 2017 from Department of Census and Statistics (2017b). Employment and output growth data for 2011 to 2015 from Department of Census and Statistics Annual Survey of Industries 2011 and 2015.

Table 3.2: Projected Labor Demand, Domestic Sector, Pessimistic and Optimistic Estimates

	2017	2018	2019	2020	2021	2022	2023	2024	2025
Pessimistic scenario									
EE	15,947	17,223	18,601	19,772	21,018	22,342	23,549	24,820	26,161
FB	261,424	287,566	316,323	336,568	358,108	381,027	399,697	419,282	439,827
Optimistic scenario									
EE	15,947	17,223	18,601	20,089	21,696	23,431	25,751	28,300	31,102
FB	261,424	287,566	316,323	344,159	374,446	407,397	451,396	500,146	554,162

EE = electronics and electricals, FB = food and beverages.
Sources: Data for 2017 from Department of Census and Statistics (2017b). Employment figures for 2018 onward based on projections using the employment elasticities set out in Table 3.1.

3.2   Estimating Skills Deficits

Next, the two-digit occupation weights based on the distribution of employment in the two subsectors in 2017 were applied to the projected employment figures for the two subsectors according to the pessimistic and optimistic scenarios set out in Table 3.2, to obtain the occupation-wise distribution of projected employment from 2018 onward. Thereafter, the subsector and occupation-specific distribution of skills data from STEP 2012 were used, as well as the occupation-specific distribution of literacy in English from LFS 2016 data as weights to derive projected skills deficits for the years 2018–2025. The skills deficits weights obtained from STEP 2012 and LFS 2016 for the two subsectors in core competencies and technical skills for high-, medium-, and low-skilled workers are set out in Table 3.3 through Table 3.7.[15] The methodology relied on a series of assumptions one of which is that the distribution of core competencies across occupations in the subsectors will remain the same as it was in 2012. The methodology and the assumptions on which it rests are set out in detail in a matrix in the Appendix (Table A.1).

---

[15]   These skills categories are defined in terms of occupation. High-skilled workers are managers, professionals, or technicians. Medium-skilled occupations include clerical workers, sales and service workers, and skilled agricultural workers. Low-skilled occupations include craft workers, machine operators and assemblers, and elementary workers.

Consider skills deficits among high-skilled workers first, as set out in Table 3.3 and Table 3.4. High-skilled workers are managers, professionals, or technicians. Decision-making skills among high-skilled workers in the FB subsector are weaker than decision-making skills of those in the EE subsector: 93% of managers in the FB subsector lack decision-making skills compared with only 27% of managers in the EE subsector. Conversely, almost all high-skilled workers in the EE subsector lack interpersonal skills while all have this skill in the FB subsector. A little more than 25% of professionals in the FB subsector cannot make presentations, while nearly a fifth in the EE subsector cannot make presentations either. In terms of technical skills, around 15% of managers in both subsectors lack supervisory and computer skills, but nearly all the high-skilled workers in the two subsectors are literate in English.

Table 3.3: STEP 2012 Weights for Deficits in Core Competencies for High-Skilled Workers in the FB and EE Subsectors

	Decision-Making Skills	Risk-Taking	Interpersonal Skills	Stability	Making Presentations
Food and Beverages					
Senior officials, managers	0.9306	1.0000	0.0000	0.0419	0.1432
Professionals	1.0000	0.8402	0.0000	0.0000	0.2803
Technicians and associate professionals	0.0000	1.0000	0.0000	0.0000	0.0000
Electronics and Electricals					
Senior officials, managers	0.2670	0.3207	0.1894	0.0000	0.1894
Professionals	1.0000	1.0000	1.0000	0.0000	0.0000
Technicians and associate professionals	0.0000	1.0000	1.0000	0.0000	0.0000

EE = electronics and electricals, FB = food and beverages, STEP = Skills Toward Employment and Productivity.
Notes: High-skilled workers are managers, professionals, or technicians. The table shows the share of workers that lack a given skill.
Source: Estimations based on World Bank (2012).

Table 3.4: STEP 2012 and LFS 2016 Weights for Deficits in Technical Skills among High-Skilled Workers in the FB and EE Subsectors

	Supervising Others	Computer Skills	Autonomy and Repetitiveness	English Literacy
Food and Beverages				
Senior officials, managers	0.1432	0.1432	0.8568	0.0332
Professionals	1.0000	0.1205	0.2803	0.0000
Technicians and associate professionals	0.0000	0.0000	0.0000	0.0243
Electronics and Electricals				
Senior officials, managers	0.1513	0.1513	0.5636	0.0000
Professionals	1.0000	0.0000	0.0000	0.0000
Technicians and associate professionals	0.0000	0.0000	0.0000	0.0854

EE = electronics and electricals, FB = food and beverages, LFS = Labour Force Survey, STEP = Skills Toward Employment and Productivity.
Notes: High-skilled workers are managers, professionals, and technicians. The table shows the share of workers that lack a given skill.
Sources: Estimations based on microdata from World Bank (2012) and Department of Census and Statistics Quarterly Labour Force Survey 2016.

Among medium-skilled occupations in the two subsectors, workers in the EE subsector appear to be lacking in several core competencies (Table 3.5). Clerical staff in the EE subsector show a lack of openness, but those who can make presentations are extroverted and show stability. In contrast, clerical workers in the FB subsector cannot make presentations but have all other core competencies. As for technical skills, clerical workers in the FB subsector show lack of computer skills, while a fourth of clerical workers in the EE subsector are similarly constrained (Table 3.6). The level of English literacy is quite good in both subsectors, with only 7% of clerical workers in EE and 6% of sales and service workers in FB unable to read and write in English. However, high scores for literacy in English may be based on respondents' ability to write their names in English, and not based on a test of functional literacy.

Table 3.5: STEP 2012 Weights for Deficits in Core Competencies among Medium-Skilled Workers in the FB and EE Subsectors

	Open	Conscientious	Extroversion	Stability	Making Presentations
Food and Beverages					
Clerical workers	0.0000	0.0000	0.0000	0.0000	1.0000
Sales and service workers and proprietors	0.1285	0.8731	0.0000	0.2364	0.8715
Skilled agricultural workers	0.0787	0.0567	0.1524	0.6375	0.2838
Electronics and Electricals					
Clerical workers	1.0000	1.0000	0.0000	0.0000	0.0000
Sales and service workers and proprietors	0.0000	0.0000	0.0000	0.0000	1.0000
Skilled agricultural workers	1.0000	1.0000	0.0000	0.0000	0.0000

EE = electronics and electricals, FB = food and beverages, STEP = Skills Toward Employment and Productivity.
Notes: Medium-skilled workers include clerical workers, sales and service workers, and skilled agricultural workers. The table shows the share of workers that lack a given skill.
Source: Estimations based on microdata from World Bank (2012).

Table 3.6: STEP 2012 and LFS 2016 Weights for Deficits in Technical Skills among Medium-Skilled Workers in the FB and EE Subsectors

	Computer Skills	English Literacy
Food and Beverages		
Clerical workers	1.0000	0.0179
Sales and service workers and proprietors	1.0000	0.0584
Skilled agricultural workers	0.9909	0.0248
Electronics and Electricals		
Clerical workers	0.2802	0.0751
Sales and service workers and proprietors	1.0000	0.0000
Skilled agricultural workers	0.0000	0.0000

EE = electronics and electricals, FB = food and beverages, LFS = Labour Force Survey, STEP = Skills Toward Employment and Productivity.
Notes: Medium-skilled workers include clerical workers, sales and service workers, and skilled agricultural workers. The table shows the share of workers that lack a given skill.
Sources: Estimations based on data from World Bank (2012) and Department of Census and Statistics Quarterly Labour Force Survey 2016.

**Table 3.7: STEP 2012 Weights for Deficits in Core Competencies
among Low-Skilled Workers in the FB and EE Subsectors**

	Core Literate	Read	Write	Numerate
Food and Beverages				
Craft and related workers	0.0415	0.3770	0.0000	0.0000
Plant and machine operators and assemblers	0.0000	0.3426	0.0000	0.0000
Elementary occupations	0.5109	0.5419	0.0875	0.0875
Electricals and Electronics				
Craft and related workers	0.1090	0.1730	0.4227	0.0864
Plant and machine operators and assemblers	0.0254	0.1226	0.0520	0.0000
Elementary occupations	0.2241	0.4208	0.7334	0.1819

EE = electronics and electricals, FB = food and beverages, STEP = Skills Toward Employment and Productivity.
Notes: Low-skilled workers include craft workers, machine operators and assemblers, and elementary workers. The table shows the share of workers that lack a given skill.
Source: Estimations based on microdata from World Bank (2012).

Tables 3.8, 3.9, and 3.10 set out the results of the estimated pessimistic and optimistic numbers of high-skilled, medium-skilled, and low-skilled workers lacking the relevant skills for the base year 2017, and for 2021 and 2025.

**Table 3.8: Projected Number of People Employed in High-Skilled Occupations
Lacking in Critical Skills in the FB and EE Subsectors**

	Baseline	Pessimistic Scenario		Optimistic Scenario	
	2017	2021	2025	2021	2025
DEFICITS IN CORE COMPETENCIES					
Food and Beverages					
Decision-making	17,674	24,210	29,735	25,315	37,465
Risk-taking	20,846	28,555	35,072	29,858	44,189
Interpersonal skills	–	–	–	–	–
Stability	458	628	771	657	972
Making presentations	3,665	5,020	6,166	5,249	7,769
Electronics and Electricals					
Decision-making	627	827	1,029	853	1,223
Risk-taking	1,204	1,587	1,976	1,639	2,349
Interpersonal skills	1,204	1,587	1,976	1,639	2,349
Stability	–	–	–	–	–
Making presentations	516	680	847	702	1,006
DEFICITS IN TECHNICAL SKILLS					
Food and Beverages					
Supervision	9,047	12,393	15,221	12,958	19,178
Autonomy	11,483	15,730	19,319	16,447	24,341
Computer skills	2,470	3,384	4,156	3,538	5,236
English literacy	451	618	759	646	956
Electronics and Electricals					
Supervision	1,728	2,277	2,834	2,351	3,370
Autonomy	415	547	681	565	810
Computer skills	111	147	183	151	217
English literacy	102	135	168	139	200

EE = electronics and electricals, FB = food and beverages.
Notes: High-skilled workers are managers, professionals, and technicians. Each individual may lack more than one competency and therefore the numbers along the columns in each subsector are not additive.
Sources: Estimations based on base employment figures generated with microdata from the Department of Census and Statistics Sri Lanka Labour Demand Survey 2017 and Quarterly Labour Force Survey 2016, and from World Bank (2012).

Table 3.9: Projected Number of People Employed in Medium-Skilled Occupations Lacking in Critical Skills in the FB and EE Subsectors

	Baseline	Pessimistic Scenario		Optimistic Scenario	
	2017	2021	2025	2021	2025
DEFICITS IN CORE COMPETENCIES					
Food and Beverages					
Openness	2,263	3,100	3,807	3,241	4,797
Conscientiousness	13,384	18,333	22,517	19,170	28,371
Extraversion	635	871	1,069	910	1,347
Stability	6,220	8,521	10,466	8,910	13,186
Making presentations	39,938	54,709	67,194	57,205	84,661
Electronics and Electricals					
Openness	634	836	1,041	863	1,238
Conscientiousness	634	836	1,041	863	1,238
Extraversion	–	–	–	–	–
Stability	–	–	–	–	–
Making presentations	516	680	847	702	1,006
DEFICITS IN TECHNICAL SKILLS					
Food and Beverages					
Computer skills	44,824	61,402	75,414	64,203	95,018
English literacy	1,442	1,976	2,427	2,066	3,057
Electronics and Electricals					
Computer skills	692	912	1,135	941	1,349
English literacy	47	62	77	64	92

EE = electronics and electricals, FB = food and beverages.
Notes: Medium-skilled workers include clerical workers, sales and service workers, and skilled agricultural workers. Each individual may lack more than one competency and therefore the numbers along the columns in each subsector are not additive.
Sources: Estimations based on base employment figures generated with microdata from the Department of Census and Statistics Sri Lanka Labour Demand Survey 2017 and Quarterly Labour Force Survey 2016, and from World Bank (2012).

Table 3.10: Projected Number of People Employed in Low-Skilled Occupations Lacking in Critical Skills in the FB and EE Subsectors

	Baseline	Pessimistic Scenario		Optimistic Scenario	
	2017	2021	2025	2021	2025
DEFICITS IN CORE COMPETENCIES					
Food and Beverages					
Core literacy	75,425	103,321	126,898	108,034	159,886
Reading	105,306	144,252	177,170	150,833	223,226
Writing	145,110	198,777	244,137	207,845	307,601
Numeracy	12,718	17,422	21,398	18,217	26,961
Electronics and Electricals					
Core literacy	435	573	714	592	848
Reading	1,577	2,079	2,588	2,146	3,077
Writing	1,123	1,481	1,843	1,528	2,191
Numeracy	140	184	229	190	273

EE = electronics and electricals, FB = food and beverages.
Notes: Low-skilled workers include craft workers, machine operators and assemblers, and elementary workers. Each individual may lack more than one competency and therefore the numbers along the columns in each subsector are not additive.
Sources: Estimations based on base employment figures generated with microdata from the Department of Census and Statistics Sri Lanka Labour Demand Survey 2017 and Quarterly Labour Force Survey 2016, and from World Bank (2012).

CHAPTER 4

THE SUPPLY OF SKILLS AND SKILLS GAPS IN THE FOOD AND BEVERAGES, AND ELECTRONICS, AND ELECTRICALS SUBSECTORS

4.1 Technical and Vocational Education and Training in the FB and EE Subsectors

This chapter examines the supply of skills in the FB and EE subsectors by reviewing skills providers and their performance in terms of graduate enrollment and output. Projections are then made of the supply of trained workers over the 2019–2025 period and matched with the estimated demand for workers projected in the previous chapter to quantify the skills gaps in each subsector. The chapter ends with a summary of the main findings and a discussion of constraints faced by the technical and vocational education and training (TVET) sector.

Public, private, and nongovernment sector training providers together provide around 40 courses for the two subsectors each year. The program mix varies by National Vocational Qualification (NVQ) status, duration, delivery mode, entry qualifications, type of service provider, and entry qualifications (Appendix, Tables A.2 and A.3). For example, skills development programs for the EE subsector provide training for low-skilled workers who have General Certificate of Education (GCE) ordinary level (OL) qualifications. The duration of low skills development programs varies from 150 hours (6 weeks) to 6 months. The entry qualification for training of medium-skilled workers is GCE advanced level (AL), and the duration varies from 1 year to 2.5 years. Most of the programs are full-time courses, and the Department of Technical Education and Training (DTET) and colleges of technology play a lead role in the development of skills for the two subsectors. For example, in the EE subsector, 50% of the programs are run by the DTET and the rest are shared by the National Apprentice and Industrial Training Authority (NAITA), Vocational Training Authority (VTA), National Youth Services Council (NYSC), and few other training providers.[16] Similarly, in the FB subsector, colleges of technology account for about 40% of course programs. In the EE subsector, the shares of the nongovernment organization (NGO) and private sectors in student enrollment is around 12% and in graduate output, 11%. In the FB subsector, their student enrollment and graduate output shares are much higher, accounting for 30% of enrollment and 38% of output.[17]

Besides the public, private, and nongovernment sector TVET institutes, several other institutions offer training facilities for the FB subsector. Among them, public sector research and extension services organizations, such as the Industrial Development Board, Industrial Technology Institute, Sri Lanka Standard Institution, and the Department of Agriculture, play a key role in conducting short-term training programs for the FB subsector. Most of these programs are subject/product-specific and are offered as 1 or 2 day workshops. For example, the Industrial Development Board offers seven

---

[16] Some of the training providers include International College of Business and Technology Ltd., SLT Campus (Pvt) Ltd., Institute of Industrial Techno Management (Pvt) Ltd., Arthur C Clarke Institute for Modern Technologies, APSS International Networks (Pvt) Ltd., Aquinas College of Higher Studies - Faculty of Engineering, etc.

[17] Some of the training providers include Prima Baking Training Centre, Sri Lanka–German Training Institute, Small Fishers Federation of Lanka, etc.

programs on food technology.[18] The Sri Lanka Standard Institution offers nine programs on quality assurance, quality management, food hygiene, food safety management, food labeling regulations, and good manufacturing practices for the food industry. There are also subsector-specific projects implemented by the donor community on food processing, baking, cookery, etc. in selected districts.[19]

More than 70% of the programs related to the FB and EE subsectors are above NVQ Level 3, and they are heavily focused on medium- and high-skilled occupation groups. This is a positive development as it ensures the supply of well-skilled workers targeted at the FB and EE subsectors (Table 4.1). However, no NVQ certificates were issued for EE subsector-related occupations in 2016 and 2017, and in the FB subsector, there was a drop in the number of NVQ certificates issued in 2017 (Table 4.2). Poor performance of NVQ assessments has been attributed to long delays in NVQ assessments (e.g., 11 months), absence of assessors, low payment rates, and long delays in making payments to assessors. It is evident that training providers have not been successful in improving skills quality and the recognition by employers of TVET graduates in the two subsectors. There is an urgent need to improve institutional efficiency in order to address these constraints.

Table 4.1: NVQ Status of TVET Programs Targeted at the FB and EE Subsectors

	FB	EE	Total	%
Non-NVQ	0	6	6	15
NVQ L3	3	2	5	12
NVQ L4	2	10	12	29
NVQ L5	3	5	8	20
NVQ L6	3	5	8	20
NVQ L7	2	0	2	5
Total	**13**	**28**	**41**	**100**

EE = electronics and electricals, FB = food and beverages, L = level, NVQ = National Vocational Qualification, TVET = technical and vocational education and training.
Source: TVEC (2018a).

Table 4.2: Status of Issuing National Vocational Qualification Certificates – 2017

FB Subsector	Level 3	Level 4	Level 5	Level 6	Total	%
Food technology	0	0	22	0	22	44
Food and beverages	26	2	0	0	28	56
Total in 2017	26	2	22	0	50	100
Total in 2016	96	2	17	0	115	

FB = food and beverages.
Source: Tertiary and Vocational Education Commission, Labour Market Information Unit.

[18]   Such as (i) instant flour (thosa / hoppers / string hoppers / idli); (ii) chocolates mixed peanut ball, peanut bar, and toffee; (iii) banana chips, manioc chips, and bite mixtures; (iv) sesame balls, sesame rolls, and sesame doshi; (v) virgin coconut oil; (vi) food packing technology; and (vii) kola kenda dried mixture, instant nutritious food mixture.

[19]   ILO, Skills for Inclusive Growth (S4IG) sponsored by the Government of Australia, and others.

About 70% of training for the FB subsector is conducted by public TVET institutes and 65% of them are targeted at medium-skilled jobs (duration of 1 year to 4 years and 3 months) while the rest are targeted at low-skilled (28%) and high-skilled (7%) occupations. The program mix of the FB subsector includes seven major programs directly relevant to the FB trade.[20] The program mix also includes a few other programs relevant to both the FB and tourism trades.[21] FB training institutes are located in the districts of Batticaloa, Colombo, Gampaha, Jaffna, Kandy, Kilinochchi, and Moneragala. The program mix of training in the EE subsector includes five major programs: National Certificate in Technology (Electrical and Electronics), National Certificate in Electronic Craft Practice (NCECP), Electronics – E2, Electronic Craftsman, Electronic Appliances Technician, Micro Controller Circuit Designing, and Higher Diploma in Electrical and Electronics Engineering. Access to training is limited to nine districts: Colombo, Galle, Gampaha, Hambantota, Jaffna, Kalutara, Kandy, Kegalle, and Matara. The higher education sector programs for the FB subsector are limited in number and are offered by public and private sector universities and other higher education institutes. The higher education sector programs for the EE subsector are also limited in number and are targeted at high-skilled occupations. The intake for university-level EE and FB programs is from GCE AL science and technology streams.

The technology stream is relatively new and is expected to attract many students for EE and FB subsector programs offered by TVET institutes, universities, and other higher education institutes. These programs are targeted at medium-skilled workers with degree or equivalent qualifications. The total number of students who passed in three subjects but are not eligible for university entry from the science and technology streams is around 40,000 per year, and the total number of students who failed all three subjects in science and technology streams is around 14,000 per year. These students are potential candidates for TVET programs in EE and FB.

4.2   Graduate Output

Total enrollment in training courses is around 1,700 students per year for the EE subsector and 2,030 students per year for the FB subsector; the TVET sector accounts for about 80% of aggregate enrollment (Figure 4.1). In terms of graduate output, however, the TVET sector accounts for 77% of graduates for the EE subsector and 94% for the FB subsector.

The EE subsector's total graduate output is distributed among low-skilled (54%) and medium-skilled (46%) occupations. Similarly, in the FB subsector, 58% of training output is accounted for by low-skilled occupations while the rest is distributed among medium-skilled (40%) and high-skilled (2%) occupations. In contrast, over 70% of the industry demand is for low-skilled occupations in both EE and FB subsectors, while medium-skilled occupations account for about 7% of total employment in the EE subsector and 17% in the FB subsector. It appears that training providers have not been able to address the issue of low demand for EE and FB subsector courses targeted at low-skilled occupations. As shown in Figure 4.2, female participation in training for the EE and FB subsectors is higher in the higher education sector relative to the TVET sector. Female enrollment is also much higher in courses related to the FB subsector than in the EE subsector.

---

[20]   Bachelor of Technology in Food Processing Technology, Higher National Diploma in Food Technology, National Diploma in Food Technology, Certificate in Fruits and Vegetable, Food Technology (NVQ Levels 5 and 6), Food and Beverage (NVQ Levels 3 and 4), Fruit and Vegetable Processor.

[21]   Baker, Cake and Cake Structure, Cook, Professional Cookery, etc.

Figure 4.1: TVET Enrollment and Graduate Output in the EE and FB Subsectors

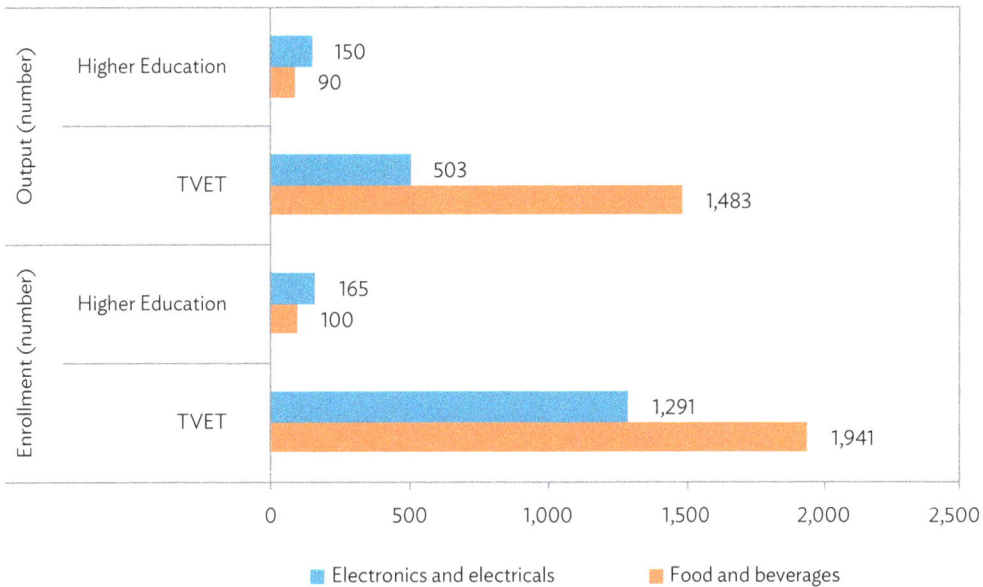

EE = electronics and electricals, FB = food and beverages, TVET = technical and vocational education and training.
Note: Average values based on enrollment and output for 2014–2017.
Sources: Tertiary and Vocational Education Commission and the University Grants Commission.

A feature that stands out in the training programs related to the two subsectors is that half of all individuals enrolled eventually drop out. Dropout rates are 51% for FB subsector courses and 61% for EE subsector courses. Women accounted for 57% of dropouts in the FB subsector and 24% in the EE subsector in 2016/2017 (Figure 4.3). The dropout rate for the EE subsector in 2016/2017 was an improvement over that of the previous year (40%), but the dropout rate for the FB subsector worsened over the same period (70%). Dalugama, Karunathilaka, and Imbulpitiya (2019) observed variations in dropout rates among students enrolled in the NCECP course at four training centers in the Western Province: Gampaha Technical College (33%), College of Technology at Maradana (33%), Ratmalana Technical College (52%), and Kalutara Technical College (27%).[22] High dropout rates in electronics have been attributed to factors such as finding a job before the training is completed, too long course duration,[23] poor understanding about job opportunities, too advanced theory components, enrollment in another course, and lack of facilities at the institution (Dalugama, Karunathilaka, and Imbulpitiya 2019). These dropout rates need to be compared with the estimated dropout rate for the entire TVET sector which is much lower at around 31%. There was also a notable difference in dropout rates between public (34%) and private (20%) training providers in the TVET sector in 2017. This suggests that public training providers are less effective in retaining students in their courses than private providers and that private provision of training may be somewhat more efficient.

---

[22] Refers to 2017 data.

[23] Duration of majority of electronics-related course programs vary from 1 year to 2 years.

Figure 4.2: Female Participation in EE and FB by Type of Service Provider, 2016/2017

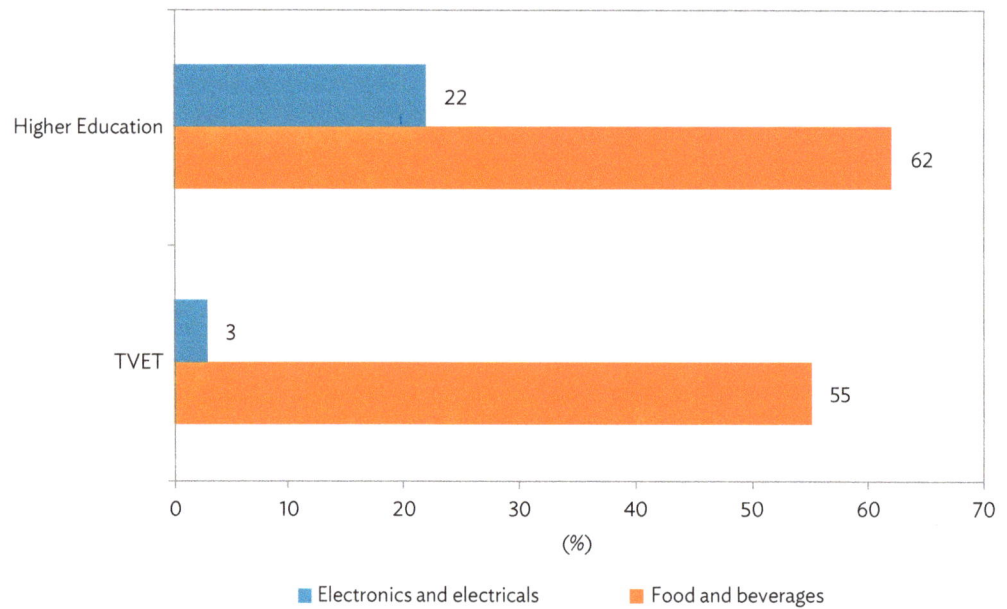

EE = electronics and electricals, FB = food and beverages, TVET = technical and vocational education and training.
Note: Average values based on enrollment and output for 2016 and 2017.
Sources: Tertiary and Vocational Education Commission and University Grants Commission.

Figure 4.3: Dropout Rates by Gender for the EE and FB Subsectors, 2016/2017

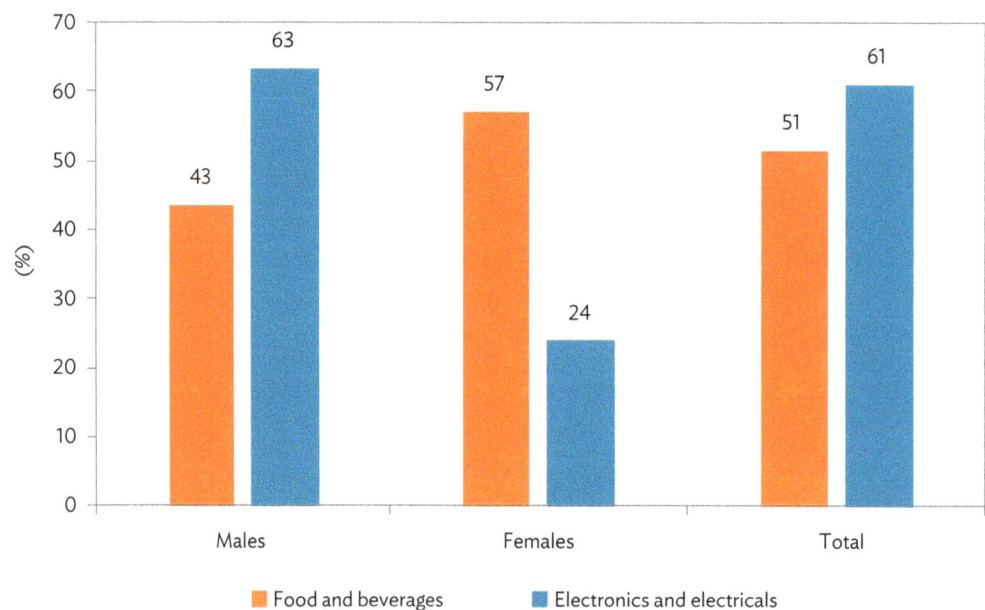

EE = electronics and electricals, FB = food and beverages.
Source: Tertiary and Vocational Education Commission, Labour Market Information Unit.

High dropout rates in TVET have been attributed to several internal and external factors. Internal factors refer to market relevance, lack of laboratory and workshop facilities, limited on-the-job training (OJT) facilities, motivation of staff and career guidance and counseling, payment of student stipends, and student welfare facilities. External factors include economic (poverty, poor health), financial (e.g., payment of fees, daily expenses, cost of travel, etc.),[24] and social factors (e.g., parents' education, low social status attached to TVET sector courses). These constraints are examined and discussed in greater detail later in this chapter.

4.3   Estimations of Graduate Output and Skills Gaps

In this section, projections are made for the graduate output of the two subsectors based on the growth performance of both TVET and higher education sector institutions that offer EE and FB subsector-related training programs. More specifically, the TVET sector graduate output estimates for the EE and FB subsectors are based on four major assumptions: (i) that average graduate output over a period of 4 years (from 2014 to 2017) will continue into later years, (ii) that the skill mix of the graduate output remains unchanged (Table 4.3), (iii) that growth rates of graduate output by type of skills (Table 4.3) remain unchanged, and (iv) that student dropout rates from 2014 to 2017 remain unchanged. The higher education sector graduate output estimates for both EE and FB subsectors are based on two premises: (i) the average graduate output over a period of 3 years (from 2015 to 2017), and (ii) an assumed annual growth rate of 5% for both subsectors. The major contributors to the EE subsector from the higher education sector include six major public universities.[25] Three major public universities contribute to building skills in the FB subsector.[26]

The projected graduate output of the EE subsector is 706 for 2019 (and is expected to increase to 947 by 2025), while that of the FB subsector is 1,766 for 2019 (expected to go up to 2,576 by 2025) (Figure 4.4). The TVET share of graduate output, which is 77% in the EE subsector and 94% in the FB subsector in 2019, is heavily concentrated in training of low-skilled workers (Figure 4.5 and Figure 4.6). By type of skills, the medium- and low-skilled categories account for 77% of graduate output in the EE subsector and 93% in the FB subsector in 2019 (Figure 4.7 and Figure 4.8). For example, about 67% of EE subsector TVET output is from four major programs of which three are 6-month non-NVQ courses: Electronic Engineering, a 6-month non-NVQ course (23%); Electronic Technology, an NVQ Level 5 course with duration of 1 year and 6 months (16%); Electronics (E2), a non-NVQ course that requires 150 hours to complete (15%); and Auto Electrical and Electronics, a non-NVQ course that requires 150 hours to complete (13%). Similarly, in the FB subsector, two major programs (Food and Beverages, and National Certificate – Fruit and Vegetable Processor) account for about 60% of graduate output and are targeted at training low-skilled workers.[27] Private and nongovernment sector training institutes account for about 12% of enrollment and 11% of graduate output of the EE subsector. In the FB subsector, they account for 30% of enrollment and 38% of graduate output.

---

[24] Currently, there is a payment of SLRs50 per day and financial assistance of SLRs2,500 per student per year for students coming from low-income families. Both private and public banks have also initiated various loan schemes to assist TVET students but the effectiveness of such schemes are yet to be seen.

[25] University of Moratuwa, University of Peradeniya, University of Jaffna, University of Ruhuna, University of Sri Jayawardenepura, and South Eastern University.

[26] University of Peradeniya, University of Sri Jayawardenepura, and Sabaragamuwa University.

[27] Both programs are at NVQ Level 3 and of 6 months' duration.

Table 4.3: Key Assumptions for Projecting FB and EE Graduate Outputs

Assumptions	Food and Beverages	Electronics and Electricals
1. Base value- TVET (number)	1,480	503
2. Base value – HE (number)	90	150
3. Growth rates –TVET (%)		
High skill	3	0
Medium skill	5	3
Low skill	7	4
4. Growth rates – HE (%)		
High skill	5	5
5. Skill mix of graduate output- TVET (%)		
High skill	2	0
Medium skill	40	30
Low skill	58	70

EE = electronics and electricals, FB = food and beverages, HE = higher education, TVET = technical and vocational education and training.
Source:　Administrative data from the Tertiary and Vocational Education Commission and the University Grants Commission.

Figure 4.4: Estimated Graduate Output in the FB and EE Subsectors

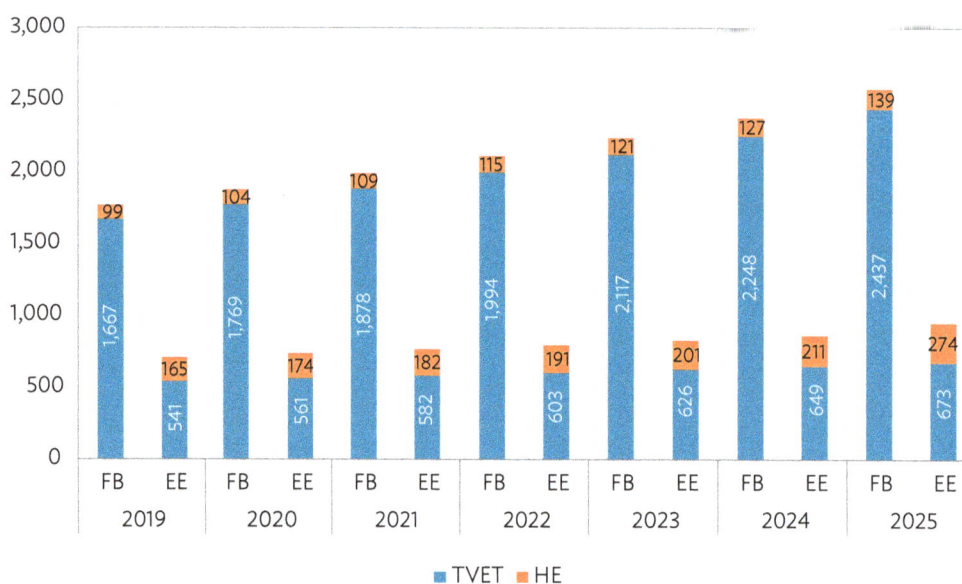

EE = electronics and electricals, FB = food and beverages, HE = higher education, TVET = technical and vocational education and training.
Source: Authors' estimates.

Figure 4.5: Projected Share of Graduate Output in the FB Subsector by Training Sector

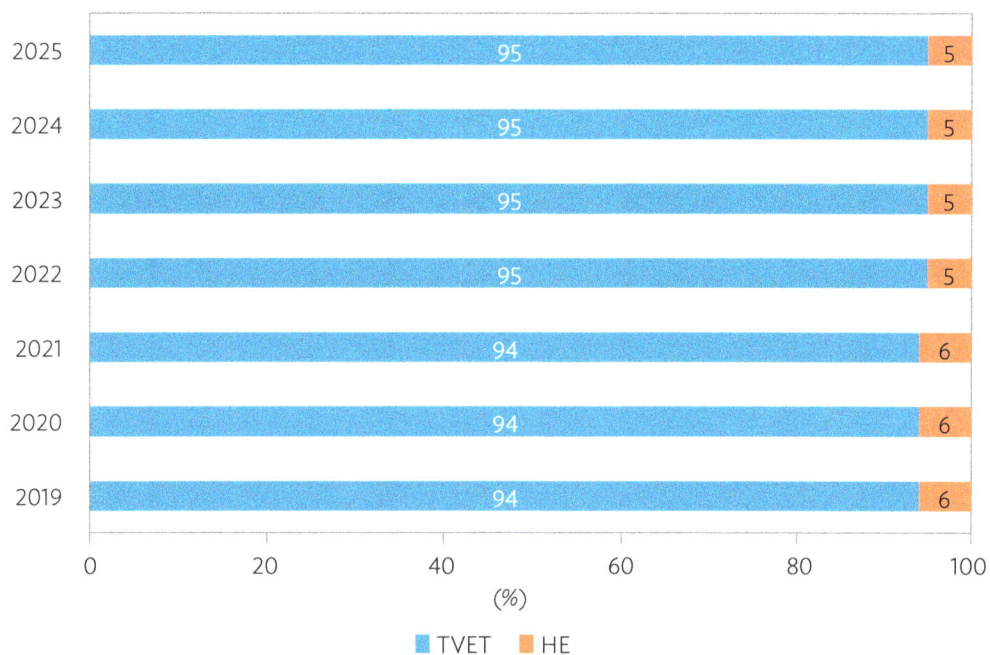

FB = food and beverages, HE = higher education, TVET = technical and vocational education and training.
Source: Authors' estimates.

Figure 4.6: Projected Share of Graduate Output in the EE Subsector by Training Sector

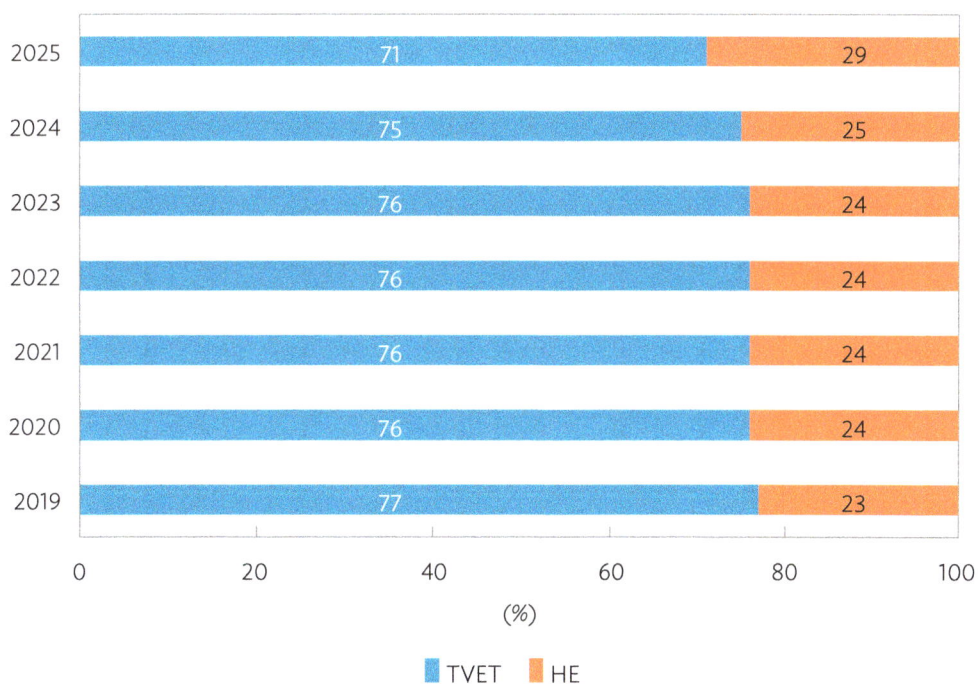

EE = electronics and electricals, HE = higher education, TVET = technical and vocational education and training.
Source: Authors' estimates.

**Figure 4.7: Projected Share of Graduate Output
in the FB Subsector by Skill Category, 2019–2025**

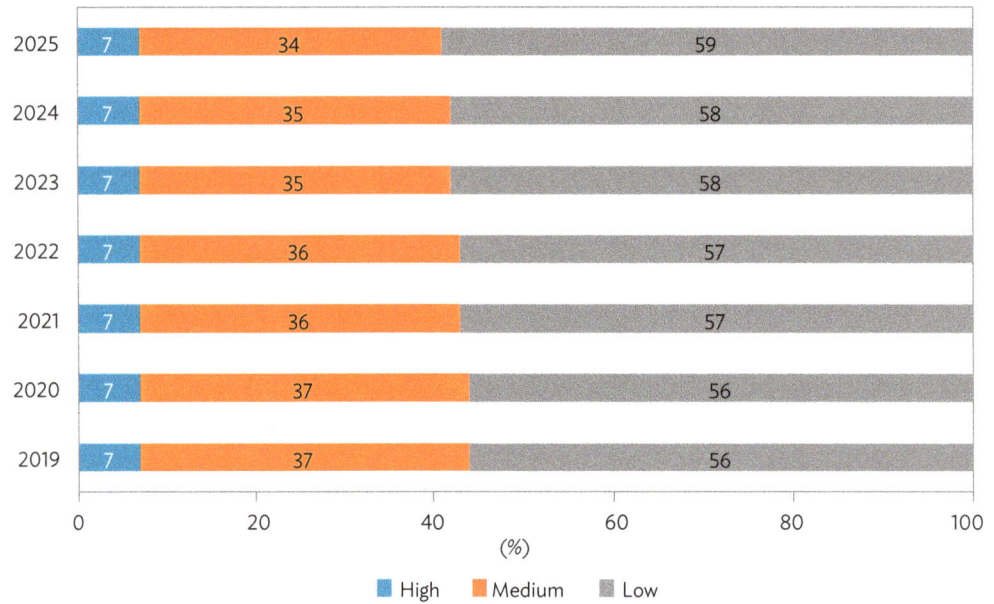

Year	High	Medium	Low
2025	7	34	59
2024	7	35	58
2023	7	35	58
2022	7	36	57
2021	7	36	57
2020	7	37	56
2019	7	37	56

(%)

■ High　■ Medium　■ Low

FB = food and beverages.
Source: Authors' estimates.

**Figure 4.8: Projected Share of Graduate Output
in the EE Subsector by Skill Category, 2019–2025**

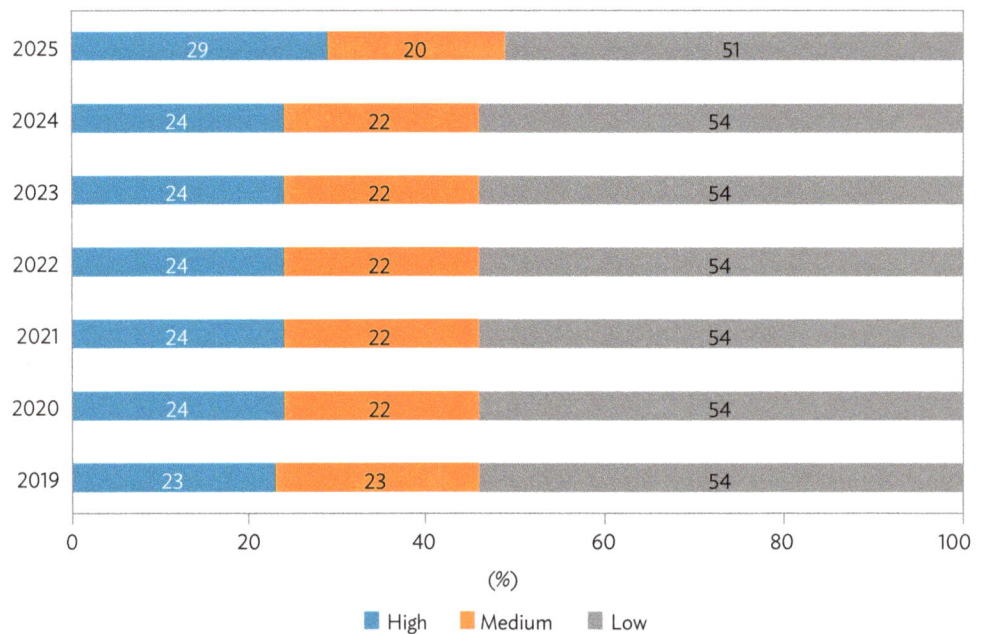

Year	High	Medium	Low
2025	29	20	51
2024	24	22	54
2023	24	22	54
2022	24	22	54
2021	24	22	54
2020	24	22	54
2019	23	23	54

(%)

■ High　■ Medium　■ Low

EE = electronics and electricals.
Source: Authors' estimates.

4.4   Skills Gaps

Skills deficits in the EE and FB subsectors are captured in the Department of Census and Statistics Labour Demand Survey of 2017 of the Department of Census and Statistics, and the data suggest that a total of 4,053 formal sector jobs are expected to be filled in the EE (59%) and FB (41%) subsectors. Estimations of skills deficits in the two subsectors by high-, medium-, and low-skilled categories using the World Bank's STEP data of 2012 are presented in Chapter 3. In this section, these estimates are matched with the supply of skills in terms of expected graduate outputs of the TVET and higher education sectors, and derived the unmet demand for skills which is called the skills gap.

Projected incremental demand for workers with the necessary skills in the FB subsector is very high: over the next 7 years (2019-2015), demand for workers is expected to average 21,752 per year under pessimistic assumptions and around 38,085 per year under optimistic assumptions. While the TVET sector's output of workers for the FB subsector is less than 10% of the total demand, even the higher-education sector's graduate output is only around 100 graduates and that is for high-skilled jobs only. Thus, graduate output can meet only 2,132 of these vacancies, leaving an unmet demand or skills gap of 19,758 per year under pessimistic assumptions during the same period. The unsatisfied demand in the FB subsector under optimistic assumptions averages 36,091 per year between 2019 and 2025 (Figure 4.9). By type of skills, the largest skills gaps in the FB subsector are expected to be in low-skilled occupations followed by medium-skilled occupations (Table 4.4).

Table 4.4: Estimated Skills Gaps in the FB and EE Subsectors by Skill Category, 2019–2025

	2019	2020	2021	2022	2023	2024	2025
FOOD AND BEVERAGES							
Pessimistic							
High	2,393	1,675	1,783	1,898	1,539	1,615	1,695
Medium	4,302	2,810	2,998	3,200	2,433	2,551	2,667
Low	20,415	14,012	14,902	15,850	12,604	13,194	13,771
Total	27,110	18,497	19,683	20,947	16,576	17,361	18,132
Optimistic							
High	2,393	2,314	2,520	2,742	3,676	4,074	4,516
Medium	4,302	4,112	4,499	4,918	6,783	7,556	8,410
Low	20,415	19,661	21,411	23,301	31,464	34,895	38,676
Total	27,110	26,087	28,430	30,961	41,923	46,525	51,603
ELECTRONICS AND ELECTRICALS							
Pessimistic							
High	368	324	342	362	337	354	371
Medium	38	21	25	28	18	21	23
Low	590	430	466	505	405	433	463
Total	996	775	833	895	760	807	857
Optimistic							
High	368	394	423	453	586	639	697
Medium	38	44	51	58	98	113	129
Low	590	652	721	795	1,189	1,333	1,492
Total	996	1,091	1,194	1,306	1,873	2,085	2,319

EE = electronics and electricals, FB = food and beverages.
Note: Estimates include graduates of technical and vocational education and training only.
Source: Authors' estimates.

Figure 4.9: Projected Skills Gaps in the FB Subsector, 2019–2025

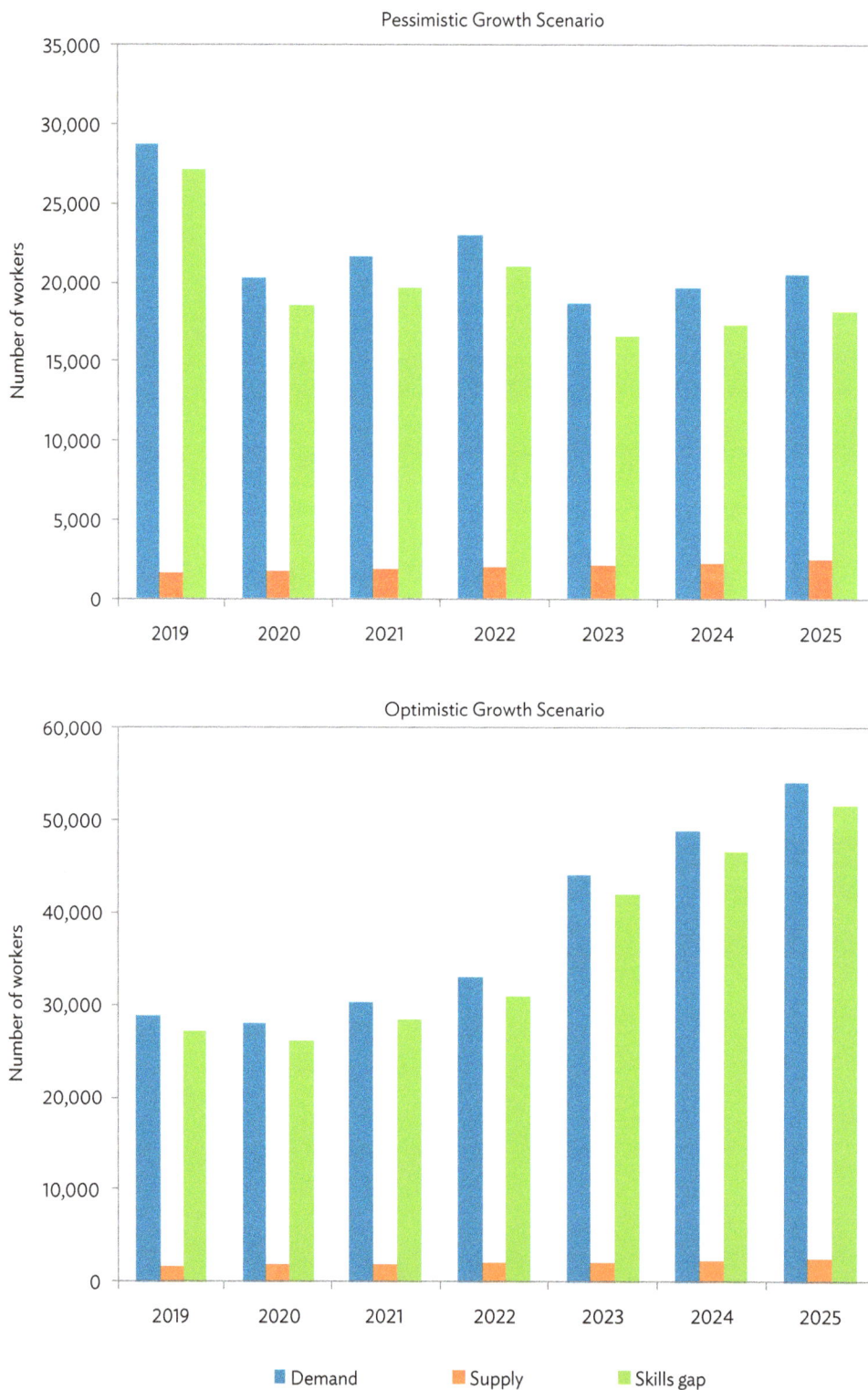

FB = food and beverages.
Source: Authors' estimates.

Figure 4.10: Projected Skills Gaps in the EE Subsector, 2019–2025

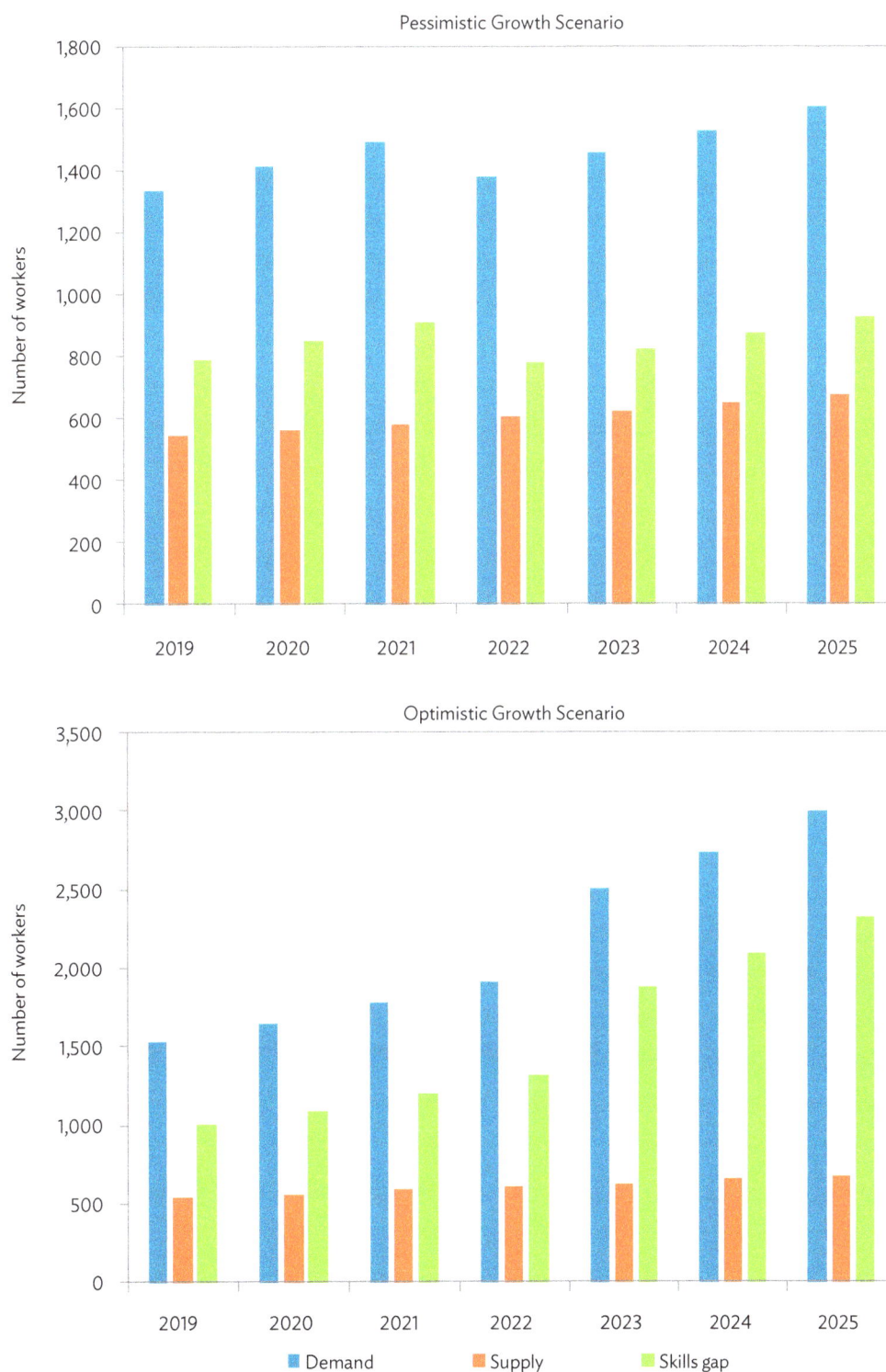

Pessimistic Growth Scenario

Optimistic Growth Scenario

EE = electronics and electricals.
Source: Authors' estimates.

In the EE subsector, the demand for workers with skills is expected to average 1,451 per year under pessimistic assumptions for the next 7 years (2019-2025).[28] The graduate outputs of the higher education and TVET sectors can meet only 800 of these projected vacancies, leaving a skills gap of 650 workers per year during the same period. Under optimistic assumptions, the unsatisfied demand would be around 2,157 per year for the next 7 years (Figure 4.10). By type of skills, the largest skills gaps in the EE subsector are expected in low-skilled occupations (e.g., 967 per year) followed by high-skilled occupations (e.g., 509 per year) (Table 4.4).

4.5   Discussion and Implications for Policy

This chapter discusses the supply side of skills development and presents the annual graduate output of TVET courses for 2019 (as being around 706 for the EE subsector and 1,766 for the FB subsector). Projected skills gaps are expected to be around 850 per year in the EE subsector and 19,758 per year in the FB subsector under pessimistic assumptions about growth over the next 7 years. These numbers will increase to 1,550 per year for the EE subsector and 36,000 per year for the FB subsector under an optimistic growth scenario. By type of skills, most of the skills gaps are among lower-skilled occupations and, hence, the TVET sector has a key role to play in reducing the severity of these skills gaps in the two subsectors.

The current supply of skills for the FB subsector meets only 10% of the total demand, whereas the supply of skills in the smaller EE subsector meets 55% of the total demand (Figure 4.9 and 4.10). Consultations with training providers on supply-side constraints revealed several factors impeding the development of skills. Among them, the lack of demand for FB and EE subsector-related courses is the most important factor that needs to be addressed immediately. Youth—particularly students—are not aware of the career opportunities in these subsectors due to the lack of national recognition for the industries. In terms of industry maturity, the FB subsector is yet to reach the stage of high value-added production with the application of new technology and high levels of human capital. Despite factor endowments and pro-market policy reforms post-1977, the FB subsector has remained an industry characterized by low technology, low value added, and weak export orientation among manufacturing subsectors. As a result, the FB subsector is not regarded as an attractive sector for employment relative to other sectors, e.g., information and communication technology (ICT), tourism, and light engineering. In contrast, the EE industry has transformed from the repair of electrical equipment to the manufacture of electrical and electronic parts, components, and equipment. With the emergence of global value chains, there has been a complete change in the structure of the EE subsector especially, and new entrants in the sector need to compete with Asian economies.

Neither the FB-related nor the EE-related course programs are impressive in terms of enrollment and graduate output between 2014 and 2017. This could be attributed to underlying factors in three key areas: (i) low demand for EE and FB programs; (ii) lack of orientation to the needs of the market; and (iii) institutional capacity constraints, particularly in responsiveness, coordination, and implementation.

The first group of constraints that TVET in the two subsectors face is the issue of low demand for courses targeted at the FB and EE subsectors. This problem has been long overlooked, and is, in fact, a key binding supply-side constraint of the FB skills training sector. It is particularly difficult for the FB subsector to attract new entrants to the labor market due to unattractive salaries and working conditions, and perceptions about its low social status. Even the training providers find it very difficult to attract enough trainees for their programs. As in most areas, employers are continually looking to hire staff who are work-ready in terms of experience and are appropriately qualified. The industry is

---

[28]   Average value of the projected total demand for 2019 to 2025 period.

looking for factory workers (food handling, pack house operator) and machine operators, low-skilled workers for the coconut and tea sectors.[29] For their part, school leavers are not aware of training facilities and employment prospects of the FB subsector. Despite high growth and labor absorptive capacity, the FB subsector receives little attention from training institutions. Thenabadu et al. (2010) reported low student participation in food-related courses and attributed the lack of interest to the lack of awareness of food trade occupations, food training opportunities, and job market demand (local and foreign). About 53% of students reported that food-related jobs lack recognition. About 50% of students also wanted to undergo FB training to get jobs abroad, while another 30% had undergone training to find their own employment. To encourage the interest of young people in agro-processing, it may be necessary to conduct communication campaigns and career guidance on the potential of the processed FB subsector. While options such as getting the necessary workers from overseas may need to be explored, as the EDB (2018b, 18) notes, "The phenomenon [skill shortages in the FB subsector] is exacerbated by the fact that operators are unable to use foreign labour."

The second issue is one of relevance. TVET programs offered for the EE and FB subsectors are out of step with industry requirements, and they need to be made more useful and relevant by introducing new technological applications and the new products industries are developing. For example, in the field of electronics, the highest demand in the industry is for industrial electronics and electronic equipment, and therefore training providers should revise course curriculum, delivery modes, and on-the-job training (OJT) facilities accordingly. More specifically, course curriculum in National Certificate in Electronic Craft Practice (NCECP) has 15 modules, but industry experts believe that some modules are not important while others may be irrelevant. Of the existing list of modules, those on magnetism, programmable logic controllers, pneumatic, electro pneumatic, microcontroller, and printed circuit board design continue to be relevant (Dalugama, Karunathilaka, and Imbulpitiya 2019).

Meanwhile, the TVET program mix of the FB subsector is narrow and needs to be expanded to cover several subsectors of the industry. The export structure of the FB industry has changed over the last 5 years, and the manufacture of processed food and animal feed products has emerged as a key subsector. Nevertheless, the lack of infrastructure necessary to expand the portfolio of courses offered constrains expansion. For example, Thenabadu et al. (2010) noted insufficient infrastructure facilities (equipment and material) at training centers, particularly in public TVET institutes, to introduce new courses. The NVQ system is also 15 years old and requires major reforms to ensure efficiency and effectiveness of quality assurance and accreditation of TVET courses. It needs to be reviewed in line with both industry demand and modern quality assurance systems of other economies, e.g., Hong Kong, China; the Philippines; the Republic of Korea; and Singapore, among others. Given the highly globalized nature of economic activities, the NVQ revisions should be aimed at promoting internationally referenced skills standard setting rather than focusing only on local demand conditions. It should also introduce more flexibility and opportunities for private sector trainers to participate in the certification process.

The issues of relevance and quality of course programs, in turn, relate to the lack of alignment of course programs with industry demand. Skills development should be demand driven and oriented toward future jobs in the labor market. Employers need to be involved in curriculum development, providing internship facilities, facilitating training of teachers, and providing post-employment training. TVET training providers also need advice and guidance from the industry on curriculum development and designing student evaluation methodologies, particularly when introducing skills-based assessments to replace traditional examination methodologies that are currently in practice. For example, an assessment of the NCECP electronic examination papers for the last 5 years has revealed that 95% of questions cover the first year's content of the curriculum which consists of basic electricity and basic electronics while material taught in the second year is not examined (Dalugama,

---

[29] Feedback responses from the Export Development Board.

Karunathilaka, and Imbulpitiya 2019). The evaluation also revealed that nearly 40% of students who sat for the examination failed in the subject of Electronics Theory. Another issue that needs to be addressed urgently is the long delays in issuing NVQ certificates. At present, it takes nearly 1 year to issue an NVQ certificate. Intra- and inter-institutional coordination needs to be improved and areas of overlapping roles need to be addressed jointly by the Department of Technical Education and Training (DTET) and the Tertiary and Vocational Education Commission (TVEC).

Limited progress in aligning EE and FB courses with technological changes in the industry is also mainly because of the lack of coordination between the industry and the TVET institutes. It is essential to recognize the need for industry participation in curriculum development, teacher training, and OJT for trainees. Thenabadu et al. (2010) and Dalugama, Karunathilaka, and Imbulpitiya (2019) also highlight the lack of collaboration between training providers and the industry in curriculum development, recognition of prior learning, OJT, and employability of trainees.

For its part, the private sector must also show firm commitment toward improving the quality and relevance of TVET programs through public–private partnership arrangements. Active private sector participation, especially from the export-oriented firms, could improve both quality and quantity of skills supply. It could improve the quality of instructors by exposing them to the most modern technological applications in industry and develop the knowledge and skills of the trainees through internship training.

This has been a perennial problem in the FB subsector. For example, Thenabadu et al. (2010) reported that about 82% of the respondents from the industry did not have strong relationships with TVET institutions, while around 78% of the respondents also indicated their willingness to collaborate with training providers. Thus, a collaborative effort by training institutes and the industry may increase student enrollment and graduate output, minimizing skills gaps in the two subsectors. Establishing separate skills councils for these two subsectors would help promote dialogue among key stakeholders.

Poor coordination of training initiatives by different public sector training providers also results in poor outcomes. While employers acknowledge the wide range of training, education, and research provision on offer across a range of public sector organizations at the central and provincial levels, they strongly recognize the need for better coordination and collaboration between training providers, industry, and policy makers. Training providers at the national level (e.g., TVEC) need to coordinate better to improve the supply of graduates and efficiency in use of resources. In the field of electronics, the DTET offers a 2-year NCECP course in electronics. The Vocational Training Authority (VTA) and National Youth Services Council (NYSC) also conduct the same NVQ standard courses with a duration of 1.5 years (VTA) and 6 months (NYSC). The VTA and NYSC also enroll students with moderate entry qualifications relative to the DTET. Another example of weak coordination is the issuance of two certificates by the DTET for its students: the DTET issues its own NCECP in addition to the NVQ certificate issued by the TVEC. This has led to excessive and undue work for students despite a nationally accepted NVQ system in place. The skills landscape in the EE and FB subsectors has become a lot more complex and, hence, coordination is essential if the two subsectors are to address skills gaps and ensure competitiveness.

Technological advances in production techniques in both industries—more so in EE—also mandate that course programs be revised and updated continuously in line with technological innovations in the industries. Moreover, instructors also need to be retrained in electronic technology and exposed to technological developments in the industry. By the standards of modern industry applications, outdated course programs on EE need to be revised in consultation with stakeholder groups from the industry. Besides, EE subsector training programs currently provided are targeted at developing skills for repair services rather than concentrating on industrial electronics. The repair service segment of the EE subsector in Sri Lanka is diminishing and, hence, training providers need to focus mainly on the supply of skills for the industrial electronics manufacturing sector.

The lack of training of trainers (TOT) has affected the quality of graduate output, resulting in low demand for TVET courses. TOT is vital for the EE and FB subsectors as the production systems of both subsectors are heavily influenced by modern technology and ICT applications. Collaboration between public sector universities and leading private sector organizations is crucial for upgrading the knowledge and skills of lecturers and instructors, and, consequently, improving the quality of FB and EE training programs. This requires effective coordination of key stakeholder groups by the TVEC.

Inadequate funding also constrains the performance of the TVET sector. Among low- and middle-income countries, Sri Lanka has the lowest spending on public education as a percentage of its GDP (e.g., 2%) as against the low- and middle-income country average of 4.5%. In the case of TVET, it is estimated that Sri Lanka allocates a total of 0.23%–0.34% of GDP (Dundar et al. 2017). TVET institutes also do not engage in income-earning activities as they are not authorized to utilize or retain earned income.

Online learning is increasingly becoming popular, and the TVET sector needs to venture into this new area of skills development at the earliest.[30] Most of the theory part of study programs could be offered online to reduce the course duration, cost of supply, and improve the standard of course modules. This could be a good strategy through which both EE and FB subsector training providers can avoid the problem of lack of qualified teachers and to make the best use of available human resources to conduct practical sessions. This would also lead to increase in enrollments due to easy access to TVET courses.

The Fourth Industrial Revolution currently underway worldwide has ushered in new digital technologies that will modify the profile of the workforce and skill requirements, and provide opportunities to increase productivity and foster industrial growth. The development of emerging technologies (e.g., industrial automation, artificial intelligence, etc.) is still at a nascent stage in Sri Lanka's manufacturing sector (ILO 2019). Although this may lead to job reductions particularly for workers who perform simple, repetitive tasks and routine cognitive work, it would also result in job gains in high-, medium-, and low-skilled occupations. For example, the demand for high-skilled jobs that require complex cognitive skills would continue to increase under the fourth wave of industrial transformation in manufacturing. Similarly, there will be job gains for medium-skilled jobs that require critical thinking and low-skilled jobs that require cooperation between humans and machines (ADB 2018).

Soft skills deficits need to be addressed at the general education level rather than overemphasizing them in courses at the TVET or higher education level. In a cross-country study, Kluve et al. (2019) observed that soft skills may not necessarily be a solution in TVET. They argue that if programs are not set up to address the needs of beneficiaries through good profiling and follow-up systems, appropriate contracting and payments systems, and a diversified package of interventions, simply adding a soft skills training component is unlikely to make a difference. Evidence from Nepal, as provided by Chakravarty et al. (2019), clearly demonstrates the positive impact of life skills in technical training on female employment and income. However, in the literature, the term "life skills" is defined broadly, ranging from writing skills, communication skills, work-readiness, life, and job search assistance, etc.[31]

---

[30] TVET sector institutes are yet to introduce online course programs. Programs even in the university sector are limited to few faculties and schools of two leading universities.

[31] For more details, see Chakravarty et al. (2019) and referenced therein.

CHAPTER 5

CONCLUSIONS AND POLICY RECOMMENDATIONS

Although Sri Lanka was the first to adopt liberal economic policies in South Asia and policy makers have repeatedly emphasized the role of export-led growth in policy documents, institutional and policy constraints have meant that Sri Lanka has not realized its full development potential through export-led growth. Key among these conditions is the provision of skills for the upgrading and diversification of the economy's productive base. Using both sample survey-based and administrative data, this study explored skills gaps in two priority manufacturing sectors which have the potential for export-led growth: the food and beverages (FB), and electronics and electricals (EE) subsectors.

This chapter presents a summary of the findings of this study about the demand for skills, the supply of skills, and the gap between them. Recommendations for policies and strategic action to address the skills gaps in the two subsectors follow.

5.1   Overview of Findings

The FB and EE subsectors together employed around 277,371 workers or 1.4% of employment and 28% of value added in manufacturing in 2017 (Table 2.11). The size distribution of firms by employment is similar in both subsectors: large (30%), medium-sized (20%), and small and micro (50%) (Figure 2.9).

Although the workforce of both subsectors is dominated by workers with only secondary education, the EE subsector workers are better educated overall. For example, 17% of workers in the FB subsector have primary (or less) education and another 53% have secondary-level education. In the EE subsector, about 30% of workers have secondary-level education and the rest have been educated at least until the GCE ordinary level. Most workers (around 92% in the FB subsector and around 71% in the EE subsector) have not undergone any post-employment training.

The projected incremental demand for workers with the necessary skills in the FB subsector is very high and, over the next 7 years, is expected to average 21,752 workers per year under pessimistic assumptions and 38,085 workers per year under optimistic assumptions. While the TVET sector's output of workers for the FB subsector is less than 10% of the total demand, even the higher education sector's graduate output is only around 100 graduates and that is for high-skilled jobs only (Figures 4.4, 4.5, and 4.6). Thus, graduate output can meet only 2,132 of these vacancies on average per year, leaving an unmet demand or skills gap of 19,640 per year under pessimistic assumptions during the same period. The unsatisfied demand in the FB subsector under optimistic assumptions averages around 37,588 per year between 2019 and 2025. By type of skills, the largest skills gaps in the FB subsector are expected to be in low-skilled occupations followed by medium-skilled occupations (Table 4.4).

In the EE subsector, the incremental demand for workers with skills is expected to average 1,451 per year under pessimistic assumptions over the next 7 years (2019-2025). The graduate outputs of the higher education and TVET sectors can meet only 800 of these projected vacancies, leaving a skills gap of 650 workers per year during the same period. Under optimistic assumptions, the unsatisfied demand averages 1,552 per year for the next 7 years. By type of skills, the largest skills gaps in the EE subsector are expected in low-skilled occupations (e.g., 967 per year) followed by high-skilled occupations (e.g., 509 per year) (Table 4.4).

Supply-side responses to increase quantity and quality of labor in both subsectors have not been sufficient. In terms of quantity of graduate output, training providers have not been able to attract enough numbers of students to conduct scheduled programs due to the low demand for EE and FB subsector courses from school leavers. School leavers with high marks for science, technology, engineering, and mathematics subjects are not interested in electronics due to a lack of awareness on job prospects, long duration of course programs, low quality of teachers, teaching of outdated course modules, and poor on-the-job training (OJT) facilities.

Consultations with training providers revealed several challenges in the TVET sector. First, updating course programs in line with industry demand seems to be a major problem due to absence of teacher training and industry exposure, poor recruitment procedures, inadequate resources, and lack of innovative changes. However, the need for drastic reforms of the entire TVET sector to meet the emerging skills demand of the industry is widely recognized. Measures such as public–private partnerships need to be looked at as a feasible option for reforming the TVET sector in line with changes in the labor market.

The government and other stakeholders need to promote digital skilling programs and strengthen core competencies of workers particularly for low-skilled occupations. Technological developments are likely to influence the task of different occupations, and training providers need to make necessary adjustments in providing occupation-specific technical skills. Digitization and automation may also affect the employment and job-specific tasks of the workforce in the two subsectors. However, due to the influence of scale economies, the effect of technology and digitization would be mainly in the large-scale segment of both subsectors which accounts for about 30% of total employment.

There is a need to improve coordination of development policies and programs between the key ministries and major institutions engaged in skills development. For example, selection of priority sectors for export-led growth at the top level should be effectively communicated to key players in the training sector so that training providers can make adjustments to their programs to increase the supply of skilled labor for priority sectors.

The general education standards in core subjects taught in the school system need urgent improvement particularly in mathematics, science, and English language. This needs to be addressed at the national level as the TVET sector cannot act as a substitute for school education.

5.2   Recommendations for Action

Well-focused subsector-specific skills development programs are urgently needed to improve productivity and output in the two subsectors as well as their ability to absorb labor. Subsector-specific skills development interventions—by addressing skills gaps both in qualitative and quantitative terms—would not only enhance industry competitiveness, but also improve efficient use of human resources at the national level. The actions proposed below arise from the findings of this study and draw on the studies by the Board of Investment of Sri Lanka, Export Development Board of Sri Lanka, and the Harvard Center for International Development (2017); and the Export Development Board (EDB) (2018a).

1. Conduct yearly campaigns in the media and in schools to create awareness among youth and young women on the potential of the EE subsectors and to encourage them to study subjects related to the subsector. For young women, in particular, the EE subsector provides good opportunities for decent work. Responsible institutions include the Ministry of Industry and Commerce, Resettlement of Protracted Displaced Persons, Co-operative Development and Vocational Training and Skills Development (MICRCDVS); TVEC; DTET; National Apprentice and Industrial Training Authority (NAITA); VTA; and Sri Lanka Electronics Manufacturers and Exporters Association (EDB 2018a, Electronic Sector, p. 45). This could be initiated and coordinated by the EDB.

2. Conduct yearly campaigns in the media and in schools to generate interest, particularly among girls, in the FB subsector on employment potential and career development of the processed FB. Since the FB subsector provides self-employment and small and medium-scale job opportunities for older workers and women, mechanisms to provide lifelong training for this cohort or those already employed in the subsector, need to be developed with FB firms as well as nongovernment organizations and microfinance providers who can act as intermediaries between the training providers and potential workers. Responsible institutions include the MICRCDVS, TVEC, DTET, NAITA, VTA, EDB, and Sri Lanka Food Processors Association (EDB 2018a, Food and Beverages Sector, p. 45).

3. Introduce an industry–educational/training institution coordination exchange mechanism by organizing short-term contracts for lecturers/instructors to serve as consultants or gain some knowledge on technological applications in the EE subsector and for experienced managers to serve as short-term, visiting lecturers to share knowledge and practical experience. Responsible institutions include the MICRCDVS, TVEC, DTET, NAITA, and VTA (EDB 2018a, Electronic Sector, p. 45). This could be initiated and coordinated by the EDB.

4. Develop short- and medium-term courses and OJT facilities through training providers, focusing on skills identified as lacking within the skills gap assessment. Responsible institutions include the MICRCDVS, TVEC, DTET, NAITA, VTA, and Sri Lanka Food Processors Association (EDB 2018a, Food and Beverages Sector, p. 45).

5. Revise course curriculum of the electronic course programs offered by training providers in consultation with the industry. Responsible institutions include the MICRCDVS, TVEC, DTET, NAITA, VTA, EDB, and Sri Lanka Electronics Manufacturers and Exporters Association. This could be initiated and coordinated by the TVEC.

6. Revise course curriculum of FB course programs offered by training providers in consultation with the industry. Responsible institutions include the MICRCDVS, TVEC, DTET, NAITA, VTA, EDB, and Sri Lanka Food Processors Association. This could be initiated and coordinated by the TVEC.

7. Conduct training-of-trainers (TOT) program for EE and FB subsectors instructors jointly by the industry and the universities. The University of Vocational Technology is expected to carry this out in addition to its study programs. Responsible institutions include the MICRCDVS, TVEC, University of Moratuwa and University of Peradeniya, EDB, Sri Lanka Food Processors Association, and Sri Lanka Electronics Manufacturers and Exporters Association. This could be initiated and coordinated by the TVEC.

8. Provide industry experience and exposure to training instructors/lecturers attached to training institutes in consultation with the industry. Responsible institutions include the

MICRCDVS, TVEC, DTET, NAITA, VTA, EDB, Sri Lanka Food Processors Association, and Sri Lanka Electronics Manufacturers and Exporters Association. This could be initiated and coordinated by the DTET and EDB jointly with private sector associations of the two subsectors.

9. Provide industry experience and exposure to trainees, particularly young women, who follow training courses offered by the training institutes. Responsible institutions include the MICRCDVS, TVEC, DTET, NAITA, VTA, EDB, Sri Lanka Food Processors Association, and Sri Lanka Electronics Manufacturers and Exporters Association. This could be initiated and coordinated by the DTET and EDB jointly with private sector associations of the two subsectors.

10. Develop occupation-specific short- and medium-term OJT courses and facilities in consultation with the industry. Developing the courses in modular formats and providing appropriate certificates recognizing the skills gained will open up pathways to acquire higher levels of skills. Responsible institutions include the MICRCDVS, TVEC, DTET, NAITA, VTA, EDB, Sri Lanka Food Processors Association, and Sri Lanka Electronics Manufacturers and Exporters Association. This could be initiated and coordinated by the DTET and EDB jointly with the private sector associations of the two subsectors.

11. Revise course evaluation methodologies of EE- and FB-related courses in consultation with the industry. At present, course evaluations are examination oriented and conducted by training institutes. Periodic tracer studies of graduates in the FB and EE subsectors must be embedded in course evaluation mechanisms. Responsible institutions include the MICRCDVS, TVEC, DTET, NAITA, VTA, EDB, Sri Lanka Food Processors Association, and Sri Lanka Electronics Manufacturers and Exporters Association. This could be initiated and coordinated by the TVEC.

12. Address the issue of long delays in issuing NVQ certificates. Responsible institutions include the MICRCDVS, TVEC, EDB, Sri Lanka Food Processors Association, and Sri Lanka Electronics Manufacturers and Exporters Association. This could be initiated and coordinated by the TVEC.

13. Encourage greater employer participation in training provision. For example, the Employment-Linked Training Purchasing model implemented by the Skills Sector Development Programme is a possible mechanism through which both the FB and EE subsectors can introduce some new course programs jointly with the industry. This could be initiated by the Sri Lanka Electronics Manufacturers and Exporters Association and the Sri Lanka Food Processors Association in consultation with the TVEC.

14. Establishment of separate skills councils for the EE and FB subsectors will help promote public–private consultation and cooperation in skills development of the two subsectors. This could be initiated by the Sri Lanka Electronics Manufacturers and Exporters Association and the Sri Lanka Food Processors Association in consultation with the MICRCDVS, TVEC, and Skills Sector Development Programme. Ensure adequate representation of women on the Councils.

15. Given that digitization and automation will affect job-specific tasks and employment, provide for digital skilling programs and embedding ICT as teaching and learning tools within TVET courses in the two subsectors. For example, creating a mobile platform to offer at least the theory part of EE and FB study programs online would be a useful mechanism that can achieve several training objectives. This could be initiated by the TVEC.

16. To improve the general education standards, specifically those related to science, technology, engineering, and mathematics subjects, there is a need to build strong foundational skills through school education by integrating 21st century knowledge and skills within the primary and secondary education curricula.

17. In view of major skills gaps in both subsectors, increased private provision of training would be a better option to meet the growing demand for skilled labor.

18. Introduction of the performance-based allowance scheme for the instructors and supervisors of the TVET sector would be an effective strategy to revamp obsolete curriculum and training courses and involve the private sector.

19. Initiate a fresh skills gap study based on a representative survey of business establishments as the existing data on the prevalence of skills is outdated (World Bank's STEP 2012) and/or has limitations (Department of Census and Statistics Labour Demand Survey 2017). This could be initiated by the MICRCDVS and TVEC in collaboration with the Department of Census and Statistics.

Finally, policy makers need to muster all available resources and expertise to reform the general education system and reverse the decline in core competencies. Without these fundamental skills in place, it will be impossible to build workplace skills at a later stage in the individual's education and training cycle. Online resources (many of the best are available free of charge) should be aggressively harnessed and made available so that young people can build a portfolio of skills that will enable them to find employment and survive, not just in these subsectors but also in the rapidly changing world of work of the future. A culture of education which promotes intellectual curiosity, ability to conceptualize and analyze problems and issues, creativity, and entrepreneurship needs to be nurtured.

APPENDIX

Figure A.1: Potential Target Groups for Skills Development

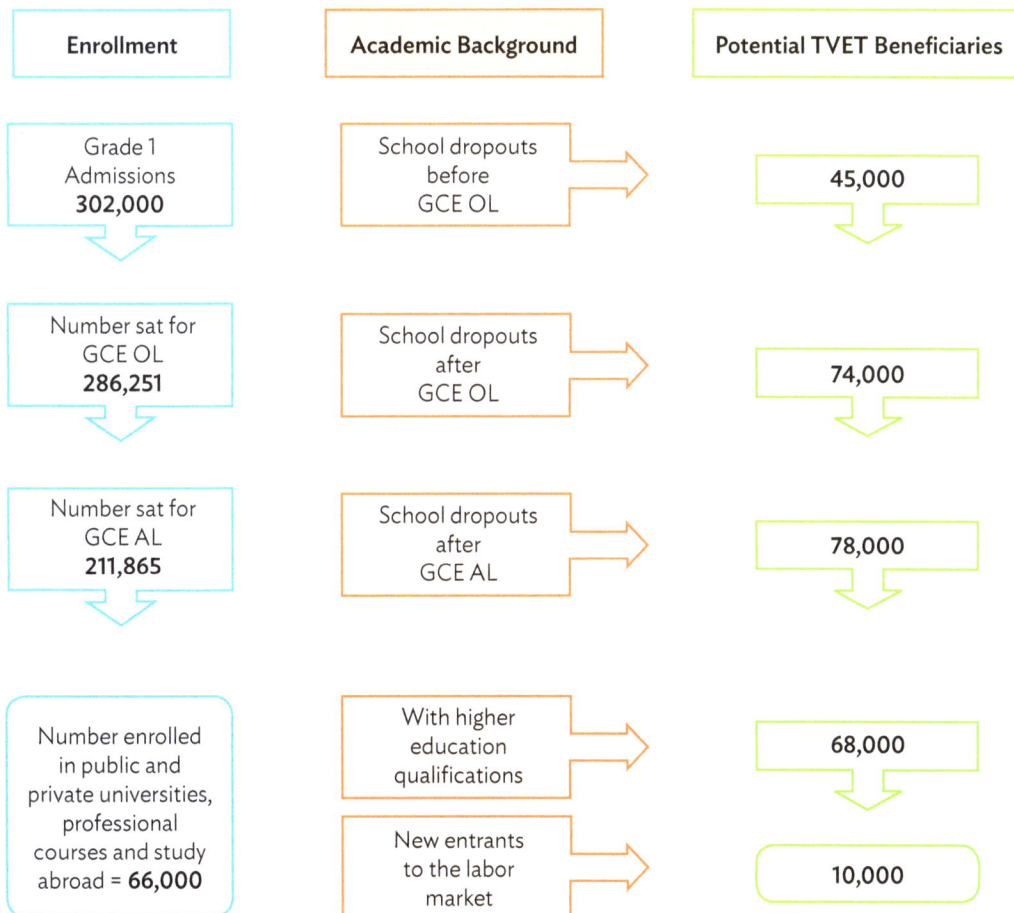

Enrollment	Academic Background	Potential TVET Beneficiaries
Grade 1 Admissions **302,000**	School dropouts before GCE OL	**45,000**
Number sat for GCE OL **286,251**	School dropouts after GCE OL	**74,000**
Number sat for GCE AL **211,865**	School dropouts after GCE AL	**78,000**
Number enrolled in public and private universities, professional courses and study abroad = **66,000**	With higher education qualifications	**68,000**
	New entrants to the labor market	**10,000**

GCE AL = General Certificate of Education advanced level, GCE OL = General Certificate of Education ordinary level, TVET = technical and vocational education and training.
Source: Authors' compilation.

**Table A.1: Methodology Used to Estimate the Demand for Skills
in Two Subsectors: Food and Beverages, and Electronics and Electricals**

Steps	Task	Data Sources	Assumptions and Remarks
1	Estimate aggregate employment figures for each subsector in 2017.	Department of Census and Statistics Labour Demand Survey 2017	
2	Derive two-digit occupation weights by estimating total employment at two-digit occupation levels in the two subsectors.	Department of Census and Statistics Labour Demand Survey 2017	
3	Estimate the distribution of cognitive, noncognitive, and technical skills at two-digit occupation level in the two subsectors. Since the number of observations for professionals and technicians occupation categories were insufficient for the food and beverages subsector, skill distribution in the manufacturing was used.	World Bank's 2012 STEP data http://microdata.worldbank.org/index.php/catalog/2017 Stata do-files on how to construct variables and aggregate measures of skills are also provided on this site and were used to estimate the skills deficits.	Assumes that the distribution of skills in 2017 is the same as it was in 2012.
4	Using estimates of the growth-related employment elasticities, project the number employed with these skills deficit structures in the two subsectors in the years 2018 to 2025.	From Step 1 to Step 3 above and estimation of growth employment elasticities	Assumptions on which the estimation of growth employment elasticities are based. Estimation of skills deficits also assumes that the skills deficit structure will remain the same over the years, and that the supply of skills will also remain the same.
5	Using estimates of the growth-related employment elasticities, estimate the number employed who lack English literacy skills in the two subsectors in the years 2016 to 2025.	Micro data from Department of Census and Statistics Labour Force Survey 2016 and the estimation of growth employment elasticities	Assumptions on which the estimation of growth employment elasticities are based. Also, assumes that the deficit structure in English literacy skills will remain the same over the years, and that the supply of skills will also remain the same.

Table A.2: List of Food and Beverages Courses Offered by the Tertiary and Vocational Education Commission, 2018

Center	Course Name (Course fee)	NVQ Level	Duration	Mode	Minimum Age	Maximum Age	Medium	Start Month
COT – Kandy	Food Technology (Free)	6	2Y 6M	Full time	17	29	E	Feb
COT – Kandy	Diploma in Food Technology (Free)	5	1Y	Full time	17	29	E	Jan
COT – Kandy	Certificate in Fruits and Vegetable Processor (Free)	3	6M	Full time	17	29	S	Jan/Jul
TC – Kuliyapitiya	Diploma in Food Technology (Free)	5	1Y	Full time	17	29	E	Jan
SLGTI – Kilinochchi	Fruit and Vegetable Processor (Free)	4	1Y	Full time	18	25	E	Sep
SLGTI – Kilinochchi	Food Technology (Free)	5	1Y	Full time	18	25	E	Sep
Batticaloa	Food and Beverage	4	3M	Full time	16	NA	S/T/E	Jan/Jul
Batticaloa	Food and Beverage	3	3M	Full time	16	NA	S/T/E	Jan/Jul
UC – Matara	HND in Food Technology	6	3Y	Full time	18	29	E	Jan
UC – Jaffna	HND in Food Technology (Free)	6	3Y	Full time	18	29	E	Apr/May
UC – Jaffna	N.Dip. in Food Technology (Free)	5	2Y	Full time	18	29	E	Apr/May
UoVT – Ratmalana	B.Tech. in Food Process Technology (SLRs60,000.00)	7	4Y 3M	Part time – weekend	NA	NA	E	Mar
UoVT – Ratmalana	B.Tech. in Food Process Technology (Free)	7	3Y	Full time	NA	NA	E	Mar
Yakkalamulla VTC	Fruit and Vegetable Processor (Free)	3	6M	Full time	16	35	S	Jan/Jul

COT = College of Technology; GCE AL = General Certificate of Education advanced level; GCE OL = General Certificate of Education ordinary level; E = English; HND = Higher National Diploma; IET = Institute of Engineering Technology; NS = not specified; NA = not applicable; NVQ = National Vocational Qualification; M = month(s); NVTI = National Vocational Training Institute; S = Sinhala; SLGTI = Sri Lanka German Training Institute; T = Tamil; TC = Technical College; UC = University College; UoVT = University of Vocational Technology; VTC = Vocational Training Centre; W = week(s); Y = year(s).
Source: TVEC (2018a).

Table A.3: List of Electronics and Electricals Courses Offered by the Tertiary and Vocational Education Commission, 2018

Institute Name	Center	Course Name (Course fee)	Entry Qualification	NVQ Level	Duration	Mode	Minimum Age	Maximum Age	Medium	Start Month
CGTTI	CGTTI – Mt. Lavinia	Electrical Maintenance – EM (SLRs16,875.00)	Others	NA	150 h	Part time	16	45	S	2 TPY
CGTTI	CGTTI – Mt. Lavinia	Electronics – E2 (SLRs16,875.00)	GCE OL	NA	150 h	Part time	16	45	S	2 TPY
CGTTI	CGTTI – Mt. Lavinia	Power Electrician (Free)	GCE OL	4	3Y 6M	Full time	16	22	S	Nov
DTET	TC – Ratmalana	NC in Engineering Craft Practice – Electronics (Free)	GCE OL	4	2Y	Full time	17	29	S	Jan
DTET	TC – Ratmalana	NC in Technology – Electrical Engineering (SLRs5,000.00)	GCE OL	5	3Y	Part time	17	NA	E	Jan
DTET	TC – Ratmalana	NC for Industrial Technician – Electrical Engineering (Free)	GCE AL	5	2Y	Full time	17	29	E	Jan
DTET	TC – Ratmalana	NC in Engineering Craft Practice – Industrial Electrician (Free)	GCE OL	4	2Y	Full time	17	29	S	Jan
DTET	TC – Ratmalana	Industrial Electronic Craftsman (Free)	Grade 9/10 Passed	4	1Y	Full time	17	29	S	Jan
DTET	TC – Kalutara	Certificate in Electrical Trade (Free)	GCE OL	3	6M	Full time	17	29	S	Jan/Jul
DTET	COT – Galle	Higher Diploma in Telecommunication Technology (Free)	GCE AL	6	2Y 6M	Full time	17	29	E	Feb
DTET	COT – Galle	Higher Diploma in Telecommunication Technology (Free)	Professional Qualifications	6	1Y	Full time	17	29	E	Feb
DTET	TC – Matara	NC in Engineering Craft Practice – Electronics (Free)	GCE OL	4	2Y	Full time	17	29	S	Jan
DTET	TC – Matara	CEB Electrical Course (Free)	GCE OL	4	1Y	Full time	17	29	E	Jan
DTET	TC – Embilipitiya	NC in Engineering Craft Practice – Industrial Electrician (Free)	GCE OL	4	2Y	Full time	17	29	S	Jan

continued on next page

Institute Name	Center	Course Name (Course fee)	Entry Qualification	NVQ Level	Duration	Mode	Minimum Age	Maximum Age	Medium	Start Month
DTET	TC – Kegalle	NC in Engineering Craft Practice – Electronics (Free)	GCE OL	4	2Y	Full time	17	29	S	Jan
DTET	COT – Kandy	NC for Industrial Technician – Electrical Engineering (Free)	GCE AL	5	2Y	Full time	17	29	E	Jan
DTET	COT – Kandy	NC in Technology – Electrical Engineering (SLRs5,000.00)	GCE OL	5	3Y	Part time	17	NA	E	Jan
NAITA	IETI – Moratuwa	Electronic Craftsman (Free)	GCE OL	4	3Y	Full time	16	35	S/E	Jan/Jun
NAITA	IET – Katunayaka	Electronics Engineering (Free)	GCE AL	6	4Y	Full time	18	25	E	Nov/Dec
NAITA	ATC – Baddegama	Electronic Appliances Technician (Free)	GCE OL	4	1Y 6M	Full time	16	35	S/E	Jan
NYSC	LAVTC – Ratmalana	Micro Controller Circuit Designing (SLRs20,000.00)	Grade 9/10 Passed	NA	6M	Part time	18	29	S	Jan/Jul
NYSC	LAVTC – Ratmalana	Micro Controller Circuit Designing (SLRs14,000.00)	Grade 9/10 Passed	NA	6M	Part time	18	29	S	Jan/Jul
NYSC	LAVTC – Ratmalana	Electronic Engineering (SLRs18,000.00)	Grade 9/10 Passed	NA	6M	Part time	18	29	S	Jan/Jul
NYSC	NYSC – Sapugaskanda	Electronic (SLRs7,050.00)	Grade 9/10 Passed	NA	6M	Part time	18	29	S	Jan/Jul
UC	UC – Ratmalana	HND in Telecommunication Technology	GCE AL	6	3Y	Full time	18	29	E	Jun
UC	UC – Ratmalana	HND in Electrical Technology	GCE AL	6	3Y	Full time	18	29	E	Jun
VTA	NVTI Orugodawatte	Electronic Technology (Free)	Professional Qualifications	5	1Y 6M	Full time	16	35	E	Jan
VTA	Dankotuwa VTC	Electronic Appliances Technician (Free)	GCE OL	3	1Y	Full time	16	35	S	Jan

ATC = Apprenticeship Training Centre; CEC = Continuing Education Centre; CEB = Ceylon Electricity Board; D = day(s); CGTTI = Ceylon-German Technical Training Institute; DTET = Department of GCE AL = General Certificate of Education advanced level; GCE OL = General Certificate of Education ordinary level; IET= Institute of Engineering Technology; IETI= Industrial Engineering Training Institute; LAVTC = Lalith Athulathmudali Vocational Training Center; M = month(s); NS = not specified; NA = not applicable; NVQ = National Vocational Qualification; NAITA = National Apprentice and Industrial Training Authority; NVTI = National Vocational Training Institute; NYSC = National Youth Services Council; TC = Technical College; TET = Technical Education and Training; TPY = terms per year; VTA = Vocational Training Authority; VTC = Vocational Training Centre; W = week(s); Y = year(s).
Source: TVEC (2018a).

REFERENCES

Asian Development Bank (ADB). 2018. *Tracer Study Sri Lanka: Public Training Institutions in 2016*. Manila.

Athukorala, P., E. Ginting, H. Hill, and U. Kumar. 2017. *The Sri Lankan Economy, Charting a New Course*. Manila: ADB.

Board of Investment of Sri Lanka, Export Development Board of Sri Lanka, and Harvard Center for International Development. 2017. Targeting Sectors for Investment and Export Promotion. https://srilanka.growthlab.cid.harvard.edu/files/sri-lanka/files/sri_lanka_ report_on_sector _targeting_exercise.pdf?m=1525360334.

Central Bank of Sri Lanka (CBSL). 2019. *Annual Report 2018*. Colombo.

Center for International Development. 2018. *Sri Lanka Growth Diagnostic*. Massachusetts: Harvard University.

Chakravarty, S., M. Lundberg, P. Nikolov, and J. Zenker. 2019. Vocational Training Programs and Youth Labor Market Outcomes: Evidence from Nepal *Journal of Development Economics*. January. 136. pp. 71–110.

Chandrasiri, S., S. De Mel, and I. Jayathunge. 2017. Enhancing Productivity and Export Competitiveness: The Case of Sri Lankan Manufacturing Industries. *Sri Lanka Economic Journal*. 14. (2). pp. 55–82.

Chandrasiri, S., and R. Gunatilaka. 2015. Skills Gaps in Four Industrial Sectors in Sri Lanka. Colombo: International Labour Organization.

Chandrasiri, S. and R. Gunatilaka. 2016. Skills for the Labor Market in Sri Lanka: An Assessment of Four Industrial Sectors. Manila: ADB. Unpublished.

Dalugama, H. K., L. K. D. M. H Karunathilaka, and V. S. Imbulpitiya. 2019. Assessing the Relevance of On the Job Training Opportunities of Radio, TV and Allied Equipment Repairer Students in DTET. Unpublished.

Department of Census and Statistics (DCS). 2012. Quarterly Labour Force Survey. Colombo.

———. 2015. *Sri Lanka Labour Force Survey, Annual Report 2014*. Colombo.

———. 2016. *Sri Lanka Labour Force Survey, Annual Report 2015*. Colombo.

———. 2017a. *Poverty Indicators*. Colombo. http://www.statistics.gov.lk/poverty/Poverty%20 Indicators_2016.pdf.

———. 2017b. Sri Lanka Labour Demand Survey, 2017. Colombo.

———. 2017c. Annual Survey of Industries, 2016. Colombo.

———. 2018. *Sri Lanka Labor Force Survey Annual Report 2017*. Colombo.

———. 2019. Annual Survey of Industries, 2017. Colombo.

Department of Jobs, Enterprise and Innovation. 2017. *Update on Future Skills Needs in the Food and Drink Sector.* 1st Report Expert Group on Future Skills Needs. Dublin, Ireland.

De Silva, W. I. 2016. *Sri Lanka: Re-emergence of Youth Bulge.* Colombo: NCAS.

De Silva, W. I., and R. de Silva. 2015. Sri Lanka: 25 Million People and Implications, Population and Housing Projections, UNFPA. Colombo: NCAS.

Dundar, H., B. Millot, M. Riboud, M. Shojo, S. Goyal, and D. Raju. 2017. *Sri Lanka Education Sector Assessment.* Washington, DC: World Bank Group.

Dundar, H., B. Millot, Y. Savchenko, H. Aturupane, and T. Piyasiri. 2014. *Building the Skills for Economic Growth and Competitiveness in Sri Lanka.* Washington, DC: World Bank Group.

Esque, T. J. and T. F. Gilbert. 1995. Making Competencies Pay Off. *Training.* 32 (1). pp. 45–50.

Export Development Board (EDB). 2017. *Export Performance Indicators of Sri Lanka, 2007-2016.*

———. 2018a. The National Export Strategy (NES) of Sri Lanka (2018–2022). Colombo: Ministry of Development Strategies and International Trade.

———. 2018b. The National Export Strategy (NES) of Sri Lanka: Processed Food and Beverage Strategy (2018–2022). Colombo: Ministry of Development Strategies and International Trade.

Gallon, M. R., H. M. Stillman, and D. Coates. 1995. Putting Core Competency Thinking into Practice. *Research Technology Management.* 38 (3). pp. 20–28.

Gunatilaka, R. 2013. Women's Participation in Sri Lanka's Labour Force: Trends, Drivers and Constraints. Colombo: ILO.

Gunewardena, D. 2015. Why Aren't Sri Lankan Women Translating Their Educational Gains into Workforce Advantages? *The 2015 ECHIDNA Global Scholars Working Paper.* Washington, DC: Centre for Universal Education at Brookings.

International Labour Organization (ILO). 2019. *Future of Work in Sri Lanka: Shaping Technology Transitions for a Bright Future.* Colombo.

Kluve, J., S. Puerto, D. A. Robalino, J. M. Romero, F. Rother, J. Stoterau, F. Weidenkaff, and M. Witte. 2019. Do Youth Employment Programs Improve Labor Market Outcomes? A Quantitative Review. *World Development.* 114 (February). pp. 237–253.

Lall, S., K. Rao, G. Wignarajah, S. D. Addario, and G. Akinci. 1996. Building Sri Lankan Competitiveness: A Strategy for Manufactured Export Growth. Report submitted to National Development Council. Colombo.

Lathi, R. K. 1999. Identifying and Integrating Individual Level and Organizational Level Core Competencies. *Journal of Business and Psychology.* 14 (1). pp. 59–75.

Majid, N., and R. Gunatilaka. 2017. The Welfare Dimensions of Employment Change in Sri Lanka and Sustainable Growth. In ADB and ILO. *Sri Lanka Fostering Workplace Skills Through Education Employment Diagnostic Study.* 2017. Manila and Geneva: ADB/ILO.

Ministry of Finance, Government of Sri Lanka. 2006. *Annual Report 2006.* Colombo.

———, Government of Sri Lanka. 2016. *Annual Report 2016.* Colombo.

Ministry of Skills Development and Vocational Training. 2017. *Progress Report – 2017.* Colombo.

National Education Research and Evaluation Centre (NEREC). 2015. *National Report: National Assessment of Students Completing Grade 8 in Year 2014 in Sri Lanka.* Colombo: Ministry of Education.

———. 2017a. *National Report: National Assessment of Students Completing Grade 8 in Year 2016 in Sri Lanka.* Colombo: Ministry of Education.

———. 2017b. *Patterns and Trends in Achievement, TIMSS: National Assessment of Students Completing Grade 8 in Year 2016 in Sri Lanka.* Colombo: Ministry of Education.

Prahalad, C. K., and G. Hamel. 1990. The Core Competence of the Corporation. *Harvard Business Review.* 68 (3). pp. 79–91.

Rama, M. 2003. The Sri Lankan Unemployment Problem Revisited. *Review of Development Economics.* 7 (3). pp. 510–525.

Tertiary and Vocational Education Commission (TVEC). 1998. *Labour Market Information Bulletin.* 1998. Colombo.

———. 2006. *Labour Market Information Bulletin,* 2006. Colombo.

———. 2011. *Labour Market Information Bulletin,* 2011. Colombo.

———. 2013. *Labour Market Information Bulletin,* 2013. Colombo.

———. 2015. *Labour Market Information Bulletin.* Vol. 2/14. December 2014, 2016. Colombo.

———. 2017a. *Labour Market Information Bulletin.* Vol. 2/16. December 2016. Colombo.

———. 2017b. *Labour Market Information Bulletin.* Vol. 2/17. December 2017. Colombo.

———. 2018a. *TVET Guide.* 2018. Ministry of Skills Development and Vocational Training, Colombo.

———. 2018b. *TVET Guide, 2017.* Colombo.

Thenabadu, M., I. Ranathunga, D. Ranasuriya, and A. Priyadarshika. 2010. Identification of Measures to Improve / Introduce Food Technology Courses in Technical and Vocational Institutes. TVEC, Research Cell. Colombo.

Weerakoon, D., U. Kumar, and R. Dime. 2019. Sri Lanka's Macroeconomic Challenges: A Tale of Two Deficits. *ADB South Asia Working Paper Series*. No. 63. Manila: ADB.

World Bank. 2011. *Transforming School Education in Sri Lanka*. World Bank, Colombo Office.

———. 2012. *Skills Toward Employment and Productivity*. Washington, DC.

———. 2015. *Sri Lanka Ending Poverty and Promoting Shared Prosperity. A Systematic Country Diagnostic*. Washington, DC: World Bank Group.

———. 2018. *Sri Lanka Development Update: More and Better Jobs for an Upper Middle-Income Country*. Washington, DC: World Bank Group.

———. 2019. *Sri Lanka Development Update: Demographic Change in Sri Lanka*. Washington, DC: World Bank Group.

www.ingramcontent.com/pod-product-compliance
Lightning Source LLC
Chambersburg PA
CBHW051657210326
41518CB00026B/2616